Experimentation

Experimentation

An Introduction
to Measurement Theory
and Experiment Design
2nd Edition

D. C. Baird

Royal Military College, Kingston, Ontario

Prentice Hall, Englewood Cliffs, NJ 07632

Library of Congress Cataloging-in-Publication Data

Baird, D. C. (David Carr)
    Experimentation.

    Bibliography: p.
    Includes index.
    1. Physical measurements. I. Title.
QC39.B17   1988   530.8   87-15515
ISBN 0-13-295338-2

Cover design: Ben Santora
Manufacturing buyer: Paula Benevento

 © 1988 by Prentice Hall
A Division of Simon & Schuster
Englewood Cliffs, New Jersey 07632

Printed in the United States of America

10 9 8 7 6 5 4 3 2 1

ISBN 0-13-295338-2   01

Prentice-Hall International (UK) Limited, *London*
Prentice-Hall of Australia Pty. Limited, *Sydney*
Prentice-Hall Canada Inc., *Toronto*
Prentice-Hall Hispanoamericana, S.A., *Mexico*
Prentice-Hall of India Private Limited, *New Delhi*
Prentice-Hall of Japan, Inc., *Tokyo*
Simon & Schuster Asia Pte. Ltd., *Singapore*
Editora Prentice-Hall do Brasil, Ltda., *Rio de Janeiro*

TO MARGARET

Contents

Preface   **xi**

1 Approach to Laboratory Work   **1**

2 Measurement and Uncertainty   **7**

2–1   Basic Nature of Measuring Process   7
2–2   Digital Display and Rounding Off   9
2–3   Absolute and Relative Uncertainty   10
2–4   Systematic Error   11
2–5   Uncertainty in Calculated Quantities   12
2–6   Uncertainty in Functions of One Variable Only   12
2–7   General Method for Uncertainty in Functions of a Single
       Variable   14
2–8   Uncertainty in Functions of Two or More Variables   16
2–9   General Method for Uncertainty in Functions of Two or More
       Variables   18
2–10   Compensating Errors   21
2–11   Significant Figures   21
       Problems   22

3 Statistics of Observation   24

3–1   Statistical Uncertainty   24
3–2   Histograms and Distributions   25
3–3   Central Values of Distributions   27
3–4   The Breadth of Distributions   29
3–5   Significance of the Mean and Standard Deviation   30
3–6   Gaussian Distribution and Sampling   31
3–7   Relation Between Gaussian Distributions and Real
       Observation   33
3–8   Sample Means and Standard Deviation of the Mean   35
3–9   Sample Standard Deviation   37
3–10   Application of Sampling Theory to Real Measurements   37
3–11   Effect of Sample Size   38
3–12   Standard Deviation of Computed Values   41
3–13   Standard Deviation of Computed Values: Special Cases   43
3–14   Combination of Different Types of Uncertainty   46
3–15   Rejection of Readings   46
       Problems   47

4 Scientific Thinking and Experimenting   49

4–1   Observations and Models   49
4–2   Construction of Models   56
4–3   Testing Theoretical Models   64
4–4   Use of Straight-Line Analysis   69
4–5   Case of Undetermined Constants   71

5 Experiment Design   76

5–1   To Test an Existing Model   76
5–2   Straight-Line Form for Equations   78
5–3   Experiment Planning   84
5–4   Experiment Design When There Is No Existing Model   89
5–5   Dimensional Analysis   90
5–6   Difference-Type Measurements   94
5–7   Experimenting with No Control over Input Variables   96
       Problems   98

6 Experiment Evaluation   102

6–1   General Approach   102
6–2   The Stages of Experiment Evaluation   104
6–3   Graphs   106
6–4   Comparison Between Existing Models and Systems   108
6–5   Calculation of Values from Straight-Line Analysis   112
6–6   Cases of Imperfect Correspondence Between System and Model   117
6–7   The Principle of Least Squares   118
6–8   Least-Squares Fit to Nonlinear Functions   121
6–9   Precautions with Least-Squares Fitting   122
6–10 Function Finding   123
6–11 Polynomial Representation   125
6–12 Overall Precision of the Experiment   126
6–13 Significant Figures   128
6–14 The Concept of Correlation   128
      Problems   133

7 Writing Scientific Reports   137

7–1   Good Writing Does Matter   137
7–2   Title   138
7–3   Format   139
7–4   Introduction   140
7–5   Procedure   142
7–6   Results   144
7–7   Graphs   146
7–8   Discussion   147

Appendices

1 Mathematical Properties of the Gaussian or Normal Distribution   151

A1–1   The Equation of the Gaussian Distribution Curve   151
A1–2   Standard Deviation of the Gaussian Distribution   155
A1–3   Areas under the Gaussian Distribution Curve   156

2 The Principle of Least Squares   158

A2–1   Least Squares and Sample Means   158
A2–2   Least-Squares Fit to Straight Lines   159
A2–3   Weighting in Statistical Calculations   161

3 Difference Tables and the Calculus of Finite Differences   163

A3–1   Mathematical Foundations   163
A3–2   Application of Difference Tables to Measured Values   169

4 Specimen Experiment   172

A4–1   Experiment Design   172
A4–2   Report   180

Bibliography   185

Solutions to Problems   187

Index   191

Preface

The first edition of this book was written to support the suggestion that, regardless of the chosen objectives for an introductory physics laboratory, the basic principles of experimenting should not be neglected and could in fact become the principal topic. Introductory laboratories in physics are particularly suited to this purpose since the systems and theories found there are usually simple enough that the basic characteristics of measurement and experimenting can easily be made visible and understandable. Such an approach to physics laboratory work can, therefore, be beneficial for a wide range of students, not only those who will proceed to professional work in physics.

That purpose on which the 1962 edition was based seems still to be valid. Many changes have taken place in the practice of experimenting, partly through the introduction of new instrumentation, but mostly because of the revolutionary impact of computing. Not only can we easily attain a level of post-experiment data analysis that would have been completely impracticable twenty-five years ago but the possibilities for the conduct of the experiment itself have been enormously expanded by the availability of on-line data analysis or computer-based control of the apparatus.

Revolutionary though such changes have been in the actual conduct of experiments, there has, nevertheless, been little or no change in the basic principles underlying the experimenting, and training in these basic principles is still required. Indeed, emphasis on these basic principles may be even more necessary today than it was twenty-five years ago on account of the present-day possibility that an experimenter can be completely insulated from the phenomena under study by an almost impenetrable barrier of data processing equipment and procedures. Under these circumstances, wholly invisible defects can produce final results with little or no meaning. Unless we have complete and clear understanding of all phases of our experiment and data analysis, we turn over our experiment wholly to the computer at our peril.

The plan of the book is largely the same as in the earlier edition but the text has been almost completely rewritten. Chapter 1 gives an outline of an approach to introductory physics laboratory work that facilitates contact with the basic nature of experimenting. Chapters 2, 3, and 4 provide the basic information on measurement, statistics and scientific procedures on which experiment design is based. Chapter 5 treats in a step-by-step way the practical requirements in designing an experiment, and Chapter 6 provides the corresponding procedures for evaluating the results of the experiment after the measurements have been made. At the end of the main text Chapter 7 contains some suggestions for writing laboratory reports.

The appendices contain material which, although desirable in itself, would have interrupted the development within the main text. This includes mathematical derivation of some of the equations quoted in the main text. In addition, a sample experiment is described in extensive detail, starting at the beginning of the experiment design, continuing through the conduct of the actual experiment and the evaluation of the results, and ending with the final report.

The material in the text has been the basis of many years of teaching in our First Year Physics Laboratory and I am grateful to the generations of students whose sometimes painful experience with it provided the opportunity for continued refinement. Finally I wish to thank Mrs. Jill Hodgson and Mme. Rachel Desrosiers for their generously-provided and invaluable assistance in preparing the manuscript.

D. C. Baird

1

Approach
to Laboratory Work

This book is intended for use in introductory physics laboratories. It was written, however, in the hope that it will serve a much wider purpose—that of providing an introduction to the study of experimenting in general, irrespective of the area in which the experimenting is carried out. Some of those studying in an introductory physics laboratory may pursue careers in physics research, and it is hoped that the book will serve as a suitable introduction to their continued studies. Many others will pursue careers in completely different areas, perhaps in other sciences. Whatever the need, the introductory physics laboratory can provide a useful introduction to the fundamental principles that underlie experimenting of any kind.

For our purposes experimentation has a very broad definition: by experimentation we mean the whole process of identifying a portion of the world around us, obtaining information from it, and interpreting that information. This definition covers a very wide range of activities—from a biologist in a white coat splicing DNA molecules to a manufacturer taking a poll to determine individual preferences in toothpaste. This book is intended to meet the needs of all who find themselves engaged in any kind of study of the world around us.

That includes those who may not themselves be actually involved in experimenting. Even if we are not actively engaged in generating it, we are all faced frequently with the requirement of at least passing judgment on experimental information offered by others. For example, our professional work may require us to make a choice between competitive bids on equipment having certain specifications, or, as members of the general public, we may be called upon to form opinions on such issues as the possible health hazards of nuclear power plants, the safety of food additives, the impact of acid rain on the environment, or the influence of national monetary policy on unemployment. What all of these examples have in common is the prominent part played by experimental information. Such public issues impose on us the responsibility of reaching our own decisions, and these decisions should be based on our assessment of the reliability of the experimental information. Even in less important matters we hear repeatedly such assertions as that scientific tests have shown that we can control our tooth decay or headaches by $x\%$ using certain products. Our choice of a new automobile may depend on our assessment of the accuracy of claimed values for fuel consumption. We are all, scientists and nonscientists alike, faced daily with the requirement to be knowledgeable concerning the nature of experimental information and the ways in which it is obtained and to be appropriately skeptical about its reliability.

To return to our claim that a physics teaching laboratory can provide an introduction to the subject of experimentation in general, it is natural to wonder how the normal laboratory with its usual experiments can be used for such a purpose. The answer lies not so much in the experiments themselves as in the attitude with which we approach them. This will become clear as our studies of experimental methods proceed, but it may be helpful at this point to illustrate the proposal with a few examples.

In offering these we must anticipate a little the work of Chap. 4 and note that we shall be viewing everything susceptible to experimental investigation in terms of "systems." By a system we mean, in general, any isolated, defined entity that functions in a specific manner. We assume that we can influence or control the system, and we refer to the methods we have available to do this as "inputs." We also assume that the system will perform some identifiable function or functions, and we refer to these as "outputs." The various examples that follow will make clear the use of the terminology. An economist, for example, may view the economy of a country as a system with an extensive set of inputs and a correspondingly varied set of outputs. The system itself will include the whole productive capacity for goods and services, transportation facilities, supply of raw materials, inhabitants, opportunities for foreign trade, weather, and many other things. The inputs are those things that can be controlled by us—the money supply, tax rates, government spending, tariffs on imports, etc. The outputs are those things that we cannot control directly; their magnitudes are determined by the system, not by us. Outputs of an economic system would include the gross national product, unemployment rate, inflation rate, external trade balance, etc. It would be very comforting and convenient if we could secure the desired values of these outputs by simple manipulation, but we cannot. No matter how desirable it may be, we cannot instruct the country's gross national product or unemployment rate to have a certain value; we are restricted to controlling our inputs. Even there we have problems. In a system as complex as a national economy the linkages between the inputs and the outputs are tangled and indirect. A change in one input variable will likely have an effect on a number of output variables, instead of the single output in which we may be interested. For example, an attempt to increase the gross national product of a country by reducing taxation rates will possibly be at least partially successful, but the simultaneous effects on other outputs may be equally prominent and not nearly as desirable—for example, a possible increase in the rate of inflation. The methods available for handling such situations are sophisticated but, with a system of this complexity, the level of success achieved by the politicians and economists shows that substantial room for improvement still remains.

There are other systems that, although still complex, are simple enough that we can control them reasonably successfully. Consider, for example, a nuclear reactor. Here the system has a smaller number of input controls and outputs, and the situation is much more clearly defined. The inputs include the position of the control rods, amount and type of fuel, rate of coolant flow, etc. The outputs include such quantities as the neutron flux density, total power produced, useful life of the fuel elements, etc. In this case the linkage between the inputs and outputs is sufficiently simple (although still not directly one-to-one) that a reasonable level of control is possible. On a more familiar level every supermarket is a system with outputs and inputs whose manipulation constitutes an experiment on the system. Every time the supermarket manager alters the price of beans (one of his inputs) he is in fact performing an experiment, for he wishes to detect a consequent change in one or more of his outputs (for example, his end-of-week profit). If he is not able to perceive the desired alteration in his outputs, he may be prompted to revise his original decision and again alter the price of beans. In other words, he is continually testing the properties of his system through experiment, and his skill in interpreting experimental results may make all the difference between profit and loss.

Incidentally, we should note in passing that, in the example of the supermarket manager, some of his inputs and outputs will involve people—work schedules, pay rates, morale, productivity, etc.—and in case this use of a systems approach to all problems, human as well as mechanical, sounds like an overmechanistic approach to life, we should note that much of our subsequent work will be concerned with limits on the validity of experimental methods. We have all heard the phrase "it has been scientifically proved" offered as an irrefutable argument, and we must be alert to the dangers of misplaced faith in scientific infallibility.

But, to return to our systems, how does all this refer to the introductory physics laboratory? In fact, if we are to prepare people to enter a scientifically literate population, would it not be better to tackle the important problems right away, and start deciding whether the mercury content of fish makes it safe to eat? The trouble is, however, that these are extremely difficult problems. Evidence is hard to obtain and its interpretation is usually uncertain; even the experts themselves disagree, often vigorously and publicly. It is almost impossible to make a significant contribution to the solution of such complex problems without first developing our skills using simpler situations. To make a start on this, let us think about some of these simpler systems.

A gasoline engine is a system that is simple in comparison with any of the earlier examples. The system includes the engine, fuel supply, mounting, surrounding atmosphere, etc. The inputs may be the obvious controls like fuel supply, fuel-air ratio, ignition timing, etc. and the outputs, as always, are the factors whose values are set by the system—the number of rpm, the amount of heat produced, the efficiency, the composition of the exhaust gases, etc. This is still a somewhat complex system, but we can begin to see that relatively simple relationships between inputs and outputs can exist. For example, the input-output relation between throttle setting and rpm for a gasoline engine is sufficiently direct and predictable that most of us invoke it daily. Note, however, that the effect of that input is not restricted to the one output in which we are interested—rpm; other outputs like heat produced, exhaust-gas composition, and efficiency are also affected by that one input, even if we are prepared, generally, to ignore the coupling.

In this example we are beginning to reach the stage at which our system is simple enough that we can start working on our theory of experimenting. Let us go one stage further and consider the example of a simple pendulum. It, too, is a system. It is, however, a system that includes very little other than the string, bob, stand, and surrounding air. Furthermore, it has only two immediately obvious inputs—the length of the string and the initial conditions according to which the motion is started. The outputs, too, are few in number. Apart from small, secondary effects, they include only the frequency of vibration and the amplitude of oscillation. Lastly, the connection between the inputs and outputs is relatively direct and reproducible. Altering the length of the pendulum's string will offer few surprises as we measure the frequency of vibration. Here, then, is a system in which the principles of experimenting will be clearly visible. If we use it to develop ability to control systems and evaluate their outputs, we shall develop the competence to tackle the more important but more complex problems later on. This gives us the key to at least some constructive uses of the introductory physics laboratory. There is real point in working with a pendulum—but only if we view it properly. If we look at it as just a pendulum, which we have all "done" before, our only reaction will be total bore-dom. If, however, we view it as a system, just like a supermarket, an airport, a nuclear reactor, or the national economy, but differing from them only in that it is simple enough that we can understand it relatively well, it will supply ex-cellent simulation of the problems of the real world.

Here we have the justification for using the introductory physics labora-tory to teach experimentation. The systems involved are sufficiently simple that they are close to being understandable, and practice with them will equip us to proceed later to our real work on complicated and important systems. We must be careful, however, about the ways in which we practice on these simple systems. We shall derive only very limited benefit if we confine ourselves to sets of instructions which tell us how to do particular experiments. If it is our intention to provide a base for proceeding to *any* type of information analysis in science, technology, business, or any of the social sciences, we shall have to provide preparation for a wide variety of experimental circumstances. In some areas random fluctuation dominates, as in the biological sciences; in oth-ers measurement may be precise, as in astronomy, but control over the subject matter is limited. The range is enormous. As we have said before, we shall try to identify general principles of experimenting, in the hope that they will be valid and useful, regardless of the future subject matter or type of experiment-ing. The remainder of this text will be concerned with those principles, and we shall assume henceforth that laboratory experiments will be regarded as exer-cises to illustrate the principles.

It may now be obvious that many of the traditional procedures in in-troductory laboratories are inappropriate for our purpose. For example, we must avoid thinking of an experiment as a procedure to reproduce some "correct" answer, deviation from which makes us "wrong." Instead, we simply assess the properties of our particular system dispassionately and take the re-sults as they come. Also, there is no point in seeking some "procedure" to fol-low: that is nothing more than asking someone else to tell us how to do the ex-periment. In real life there is rarely someone waiting to tell us what to do or what our result should be; our usefulness will depend on our ability to make our *own* decisions about handling the situation. It takes a great deal of practice and experience to develop confidence in our own decisions about the conduct of any experimental procedure, and the elementary laboratory is not too early to start. We shall, therefore, place a great deal of emphasis on experiment planning, for this is the stage at which much of the skill in experimenting is needed. It is important to avoid the temptation to regard preliminary planning as a waste of time or a distraction from the supposedly more important task of making the measurements. Time must be explicitly set aside for adequate anal-ysis and planning of the experiment before a start is made on the actual mea-suring process.

It is necessary, also, to learn to work within the framework of the apparatus available. All professional experimenting is subject to limits on resources, and much of the skill in experimenting lies in optimizing the experimental yield from these resources. Restrictions on time, too, merely simulate the circumstances of most actual experimenting. The apparatus itself will never be ideal. This should not, however, be regarded as a defect but as a challenge. The real work of evaluating experimental results consists of separating the grain of useful results from the chaff of error and uncertainty. The experimenter must learn to identify sources of error or uncertainty for himself, and, if possible, eliminate them or correct for them. Even with the greatest care, however, there will always be an irreducible residuum of uncertainty, and it is the experimenter's responsibility to evaluate the precision of the final answer, a quantity which is just as important as the answer itself. The ability to cope with such requirements can be acquired only by actual contact with realistic working conditions, and it is a common injustice to students in introductory physics laboratories to provide apparatus that is too-carefully adjusted, or to give, in other ways, the impression that the experiments are ideal. This is unfortunate, because the foundation of future expertise lies in constructive response to experimental limitations.

In summary, the most fruitful use of laboratory time will result when the experiments are accepted as problems to be solved by the student himself. Certainly, errors in judgment will be made, but we can learn more effectively from personal experience of the consequences of our decisions than we can by following rigidly some established, "correct" procedure. What we learn is more important than what we do. This is not to say, however, that we should show complacent indifference to the outcome of the experiment. Development of our experimenting skills will come about only if we take seriously the challenge of obtaining the best possible result in every experiment.

The writing of laboratory reports should be tackled in the same constructive spirit. In professional life there is very little point in spending time and trouble on an experiment unless we can adequately convey the outcome to others. We have an obligation to our readers to express ourselves as lucidly, if not elegantly, as possible. It is wrong to regard this as the responsibility of departments of English, and report writing in the introductory science laboratory should be accepted as an opportunity for exercise in descriptive composition. Report writing that degenerates into a mere indication that the experiment has been performed is little more than a waste of time and a loss of opportunity for necessary practice. Report writing at the level suggested here is almost pointless without adequate review and criticism. Opportunities for improvement become much more obvious in hindsight, and such detailed review should be regarded as an indispensable part of the work in a teaching laboratory.

2

Measurement and Uncertainty

2–1 BASIC NATURE OF MEASURING PROCESS

Measurement is the process of quantifying our experience of the external world. The nineteenth-century Scottish scientist, Lord Kelvin, once said that "when you can measure what you are speaking about and express it in numbers, you know something about it; but, when you cannot measure it, when you cannot express it in numbers, your knowledge is of a meager and unsatisfactory kind; it may be the beginning of knowledge, but you have scarcely in your thoughts advanced to the stage of science." While this may be a slight overstatement, it remains true that measurements constitute one of the basic ingredients of experimenting. We shall not reach a satisfactory level of competence in experimenting without knowledge of the nature of measurement and the significance of measurement statements.

It is obvious that the quantifying process will almost invariably involve comparison with some reference quantity (how many paces wide is my back yard?). It is equally obvious that the good order of society requires extensive agreement about the choice of reference quantities. The question of such measurement standards, defined by legislation and subject to international agreement, is extensive and important. No one seriously interested in measurement can ignore the question of defining and realizing standards in his area of work. A discussion of this important topic here would, however, distract us from our chief concern, the process of measuring. We shall, therefore, leave the topic of standards without further mention except reference to the texts listed in the Bibliography, and take up the study of actual measuring processes.

Let us start at the most basic level with an apparently simple measurement; let us try to find out what kind of process is involved and what kind of statement can be made. If I give the notebook in which this is being written to someone and ask him to measure its length with a meter stick, the answer is absolutely invariable—the length of the notebook is 29.5 cm. But that answer must make us wonder: are we really being asked to believe that the length of the book is exactly 29.50000000 . . . . cm? Surely not; such a claim is clearly beyond the bounds of credibility. So how are we to interpret the answer? A moment's thought in the presence of the notebook and a meter stick will make us realize that, far from determining the "right" or "exact" value, the only thing we can realistically do is approach the edge of the notebook along the scale, saying to ourselves as we go: "Am I sure the answer lies below 30 cm? Below 29.9 cm? Below 29.8 cm?" The answer to each of these questions will undoubtedly be "Yes." As we progress along the scale, however, we shall eventually reach a point at which we can no longer give the same confident reply. At that point we must stop, and we identify thereby one end of an interval that will become our measured value. In a similar way we can approach the edge of the notebook from below, asking ourselves at each stage: "Am I sure that the answer lies above 29.0 cm? 29.1 cm," and so on. Once again we shall reach a value at which we must stop, because we can no longer say with confidence that the answer lies above it. By the combination of these two processes we identify an interval along the scale. It is the smallest interval that, as far as we can be certain, does contain our desired value; within the interval, however, we do not know where our answer lies. Such is the only realistic outcome of a measuring process. We cannot look for exact answers, and we must be content with measured values that take the form of intervals. Not only does this example illustrate the essential nature of a measuring process, it also provides guidance for actually making measurements. The process of approaching the value we seek from each side separately reminds us of the necessity of stating the result as an interval, and also makes it easier to identify the edges of that interval.

The final outcome of our discussion is a most important one. As we make measurements and as we report the results we must keep in mind constantly this fundamental and vital point—measurements are not exact, single numbers but consist of intervals, within which we are confident that our desired value lies. The act of measurement requires us to determine both the location and width of this interval, and we do it by the careful exercise of visual judgment every time we make a measurement. There are no rules for determining the size of the interval, for it will depend on many factors in the measuring process. The type of measurement, the fineness of the scale, our visual acuity, the lighting conditions—all will play a part in determining the width of the measurement interval. The width, therefore, must be determined explicitly each time a measurement is made. For example, it is a common error to believe that, when making a measurement using a divided scale, the "reading error" is automatically one half of the finest scale division. This is an erroneous over-simplification of the situation. A finely divided scale used to measure an object with ill-defined edges can give a measurement interval as large as several of the finest scale divisions; a well-defined object and good viewing conditions, on the other hand, may permit the identification of a measurement interval well within the finest scale division. Every situation must be assessed individually.

2–2 DIGITAL DISPLAY AND ROUNDING OFF

Other aspects may also confuse the issue. Consider, for example, a piece of equipment which gives a digital readout. If a digital voltmeter tells us that a certain potential difference is 15.4 V, does it intend to imply that the value is 15.40000 . . . exactly? Clearly not, but what does it mean? That depends on circumstances. If the instrument is made in such a way that it reads 15.4 V because the actual value is closer to 15.4 than it is to 15.3 or 15.5, then the meaning is: this reading lies between 15.35 and 15.45. On the other hand, a digital clock may be made in such a way that it changes its indication from 09.00 to 09.01 at the time of 09.01. When we see it reading 09.00, then, we know that the time lies between 09.00 and 09.01, a slightly different interpretation from that appropriate to the digital voltmeter. Again, each situation must be judged by itself.

These two examples of digital display illustrate a more general concept, the inaccuracy inherent in the process of "rounding off." Even without inaccuracy arising from limited ability to make measurements, a mere statement of a numerical quantity can contain inaccuracy. Consider the statement

$$\pi = 3.14$$

We all know that this is not so because we can remember some, at least, of the following numbers, 3.14159 . . . . So what can we mean by quoting π as 3.14? It can mean only that π has a value closer to 3.14 than it does to 3.13 or 3.15. Our statement is, therefore, that π lies between 3.135 and 3.145. This range of possibility represents what is sometimes known as a "rounding-off error." Such errors can be small and unimportant, or they can become significant. In a long calculation, for example, there is a chance that rounding-off errors can accumulate, and it becomes wise, especially in these days of conveniently available calculators, to carry through the calculation more figures than one might think would be necessary. A similar rounding-off error can appear in statements about measurement. We sometimes hear that someone has made a measurement on a scale which was "read to the nearest millimeter" or some such phrase. This is not a very good way of reporting a measurement because it obscures the actual value of the measurement interval. We do, however, encounter such statements and, if we are obliged to deal with a measurement quoted in that form, we can only assume that the scale division quoted represents some kind of minimum value for the size of the measurement interval.

2–3 ABSOLUTE AND RELATIVE UNCERTAINTY

By whatever means we have made a measurement, the final outcome should be an interval which represents, to the best of our ability, the range inside which the desired value lies. In the example we used first the experimenter might be able to state with confidence no more than that the length of the notebook lay between 29.4 and 29.6 cm. Although the only meaningful outcome of a measuring process consists of such an interval or range, it is frequently desirable, for purposes of description or further calculation, to rephrase the quoted value. We take the interval 29.4–29.6 and *rename* it 29.5 ± 0.1 cm. Although obviously no more than a renamed expression of the original interval, the new form does offer certain advantages. It gives us a central value, 29.5, which can be used in further calculations. It also gives us a value, ±0.1, called the "uncertainty" of the measurement, by which we can judge the quality of the measuring process and which can be used in separate calculations on uncertainties. One disadvantage in this mode of expression is the return to a central value, 29.5. Unless we remember clearly that only the complete quantity, 29.5 ± 0.1, serves as an adequate statement of the answer, we may become sloppy in making and reporting measurements and may forget the essential presence of the uncertainty. We should all make it an invariable practice to associate an uncertainty value with a reading, both at the time we make the measurement, and subsequently, whenever the value is quoted or used in further calculation.

Since the figure ±0.1 cm represents the actual amount, or range, by which the reading of 29.5 is uncertain, it is often called the "absolute uncertainty" of the reading, and we shall consistently use this terminology. In addition, other aspects soon become important. How significant is an uncertainty of ±0.1 cm? When we are measuring the length of a notebook, it is significant to a certain extent. If we are measuring the distance between two cities, an uncertainty of ±0.1 cm is probably completely insignificant. If, on the other hand, we are measuring the size of a microscopic bacterium, an uncertainty of

±0.1 cm would make the measurement meaningless. For this reason, it is frequently desirable to compare an uncertainty figure with the actual value of the measurement; by so doing the significance of the uncertainty can be realistically assessed. We define the ratio

$$\textbf{relative uncertainty} = \frac{\textbf{absolute uncertainty}}{\textbf{measured value}}$$

In the case of our example

$$\text{relative uncertainty} = \pm\,\frac{0.1}{29.5} = \pm 0.003$$

This relative uncertainty is often quoted as a percentage, so that, in the present case, the relative uncertainty would be ±0.3%. Such a quantity gives us a much better feeling for the quality of the reading, and we often call it the "precision" of the measurement. Note that the absolute uncertainty has the same dimensions and units as the basic measurement (29.5 cm is uncertain by 0.1 cm), while the relative uncertainty, being a ratio, has no dimensions or units and is a pure number.

2–4 SYSTEMATIC ERROR

The kind of uncertainty that we have been considering arises from naturally occurring inadequacy in the measuring process. A different type of error can appear when something affects all the measurements of a series in an equal or a consistent way. For example, a voltmeter or a micrometer caliper can have a zero error, a wooden meter stick may have shrunk, a person may consistently press a stopwatch button $\frac{1}{10}$ sec behind the event, and so on. These errors are termed "systematic errors," a subclass of which are "calibration errors." Because such systematic errors may not be immediately visible as one makes a measurement, it is necessary to be vigilant and remember at all times the possibility of their presence. Instrumental zeroes, for example, should automatically be checked every time an instrument is used. Although it may be less easy to check calibration, the accuracy of electrical meters, stopwatches, thermometers, and other such instruments should not be taken for granted and should be checked whenever possible. Also, the presence on an instrument of a precise-looking, digital readout with four or five supposedly significant figures should not be taken as proof of precision and freedom from systematic error. Most of a batch of electronic timers that our laboratory recently acquired for laboratory teaching, which could supposedly measure time intervals with millisecond precision, turned out to have calibration errors as large as 14%. Do not be deceived; view all measuring instruments with suspicion and check instrumental calibration whenever possible.

2–5 UNCERTAINTY IN CALCULATED QUANTITIES

In the preceding sections we have been concerned solely with the concept of uncertainty in a single measurement. It is rare, however, that a single measurement ends the process. Almost invariably the result we desire is a combination of two or more measured quantities or is, at least, a calculated function of a single measurement. We might wish, for example, to calculate the cross-sectional area of a cylinder from a measurement of its diameter, or its volume from measurements of both diameter and length. The various measurements will sometimes be of different types, as in a calculation of g from values of the length and period of a pendulum. In these cases the presence of uncertainty in the basic measurements will obviously entail the presence of uncertainty in the final computed value. It is this final uncertainty that we now wish to calculate. For the purposes of this section we shall assume that our uncertainties have the character of ranges or intervals within which we are "almost certain" that our answer lies. For the computed values we shall calculate intervals within which we wish, once again, to be "almost certain" that our answer lies. That means that we must do our calculation for the "worst case" of combined uncertainties. This is perhaps a pessimistic assumption, and we shall see later, in Chap. 3, how the probabilities associated with various error combinations enable us to make a more realistic and less pessimistic estimate. For the moment, however, let us assume that we wish to calculate, from the uncertainties in the primary values, the maximum range of possibility for the computed answer.

2–6 UNCERTAINTY IN FUNCTIONS OF ONE VARIABLE ONLY

Consider a measured quantity x_0 with an uncertainty $\pm \delta x$, and consider a computed result z to be a function of the variable x. Let

$$z = f(x)$$

This function enables us to calculate the required value z_0 from a measured value x_0. Moreover, the possibility that x can range from $x_0 - \delta x$ to $x_0 + \delta x$ implies a range of possible values of z from $z_0 - \delta z$ to $z_0 + \delta z$. We now wish to calculate the value of δz. The situation is illustrated graphically in Fig. 2–1, in which, for a given $f(x)$, we can see how the measured value x_0 gives rise to the computed result z_0, and how the range $\pm \delta x$ about x_0 produces a corresponding range $\pm \delta z$ about z_0.

Before considering general methods of evaluating δz it is instructive to see how finite perturbations are propagated in simple functions. Consider, for example, the function

$$z = x^2$$

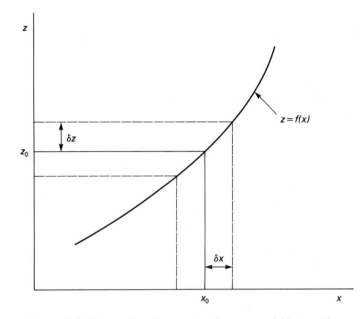

Figure 2–1   Propagation of uncertainty from one variable to another.

If x can range between $x_0 - \delta x$ and $x_0 + \delta x$, then z can range between $z_0 - \delta z$ and $z_0 + \delta z$, where

$$z_0 \pm \delta z = (x_0 \pm \delta x)^2$$
$$= x_0^2 \pm 2x_0\delta x + (\delta x)^2$$

We can ignore $(\delta x)^2$, since δx is assumed to be small compared with x_0, and equate z_0 to x_0^2, giving for the value of δz

$$\delta z = 2x_0\delta x$$

This can more conveniently be expressed in terms of the relative uncertainty $\delta z / z_0$:

$$\frac{\delta z}{z_0} = \frac{2x_0\delta x}{x_0^2} = 2\frac{\delta x}{x_0}$$

Thus, the relative uncertainty of the computed result is twice that of the initial measurement.

Although it is essential to bear in mind the nature of propagated uncertainty, as illustrated by the use of finite differences, considerable simplification of the formulation can be achieved using differential calculus.

2-7 GENERAL METHOD FOR UNCERTAINTY IN FUNCTIONS OF A SINGLE VARIABLE

In the preceding section the finite differences δz and δx are merely an expression of the derivative dz/dx. We can therefore obtain our value of δz by first using standard techniques to obtain dz/dx in the form

$$\frac{dz}{dx} = \frac{d(f(x))}{dx}$$

and then writing

$$\delta z = \frac{d(f(x))}{dx}\, \delta x \qquad\qquad (2\text{-}1)$$

This is a relatively simple procedure, and it will work well in cases for which the elementary, finite-difference approach would lead to algebraic complexity. Thus, if

$$z = \frac{x}{(x^2 + 1)}$$

then

$$\frac{dz}{dx} = \frac{x^2 + 1 - x \cdot 2x}{(x^2 + 1)^2}$$

$$= \frac{1 - x^2}{(1 + x^2)^2}$$

and

$$\delta z = \frac{1 - x^2}{(1 + x^2)^2}\, \delta x$$

This calculation would have been very awkward by any other approach. Furthermore, it gives δz generally as a function of x and δx; any particular desired value can be obtained by setting $x = x_0$. Let us now use this technique to evaluate uncertainties for some common functions.

(a) Powers

Consider

$$z = x^n$$

$$\frac{dz}{dx} = nx^{n-1}$$

$$\delta z = nx^{n-1}\delta x$$

The significance of this result becomes a little more obvious when expressed in terms of the relative uncertainty. Thus,

$$\frac{\delta z}{z} = n\frac{\delta x}{x}$$

Thus, when evaluating powers, the *relative uncertainty* of the result is the relative uncertainty of the basic quantity multiplied by the power involved. This will be valid for either powers or roots, so that precision diminishes as a quantity is raised to powers and improves on taking roots. This situation must be carefully watched in an experiment in which powers are involved. The higher the power, the greater will be the need for high initial precision.

(b) Trigonometric Functions

We shall do only one example, since all the others can be treated in similar ways. Consider

$$z = \sin x$$

Here

$$\frac{dz}{dx} = \cos x$$

and

$$\delta z = (\cos x)\, \delta x$$

This is one case where the elementary method of inserting $x_0 \pm \delta x$ shows the result more clearly. Using the approximation

$$\cos \delta x = 1$$

we obtain

$$\delta z = \cos x \sin \delta x$$

showing that the δx in the preceding result is really $\sin \delta x$ in the limit of small angles. Only in the case of very large uncertainty would this difference be significant, but it is best to understand the nature of the result. Clearly δx should be expressed in radian measure. The result will normally have straightforward application when dealing with apparatus such as spectrometers.

(c) Logarithmic and Exponential Functions

Consider

$$z = \log x$$

Here

$$\frac{dz}{dx} = \frac{1}{x}$$

and

$$\delta z = \frac{1}{x}\delta x$$

The relative uncertainty can be calculated as usual. If

$$z = e^x$$

$$\frac{dz}{dx} = e^x \; .$$

and

$$\delta z = e^x \delta x$$

This is an important case, since exponential functions occur frequently in science and engineering. These functions can become very sensitive to the exponent when it takes values much over unity, and the uncertainty δz may become very large. This will be familiar, for example, to anyone who has watched the current fluctuations in a thermionic diode that can result from quite small variations in filament temperature.

As stated earlier, the method can be easily applied to any function not listed above by evaluating the appropriate derivative and using Eq. (2–1).

2–8 UNCERTAINTY IN FUNCTIONS OF TWO OR MORE VARIABLES

If the result is to be computed from two or more measured quantities, x, y, etc., the uncertainty in the result can, as was mentioned in Sec. 2–5, be regarded in two different ways. We could be as pessimistic as possible and suppose that the actual deviations of x and y happen to combine in such a way as to drive the value of z as far as possible from the central value. We would, in this way, calculate a value for δz which gives the extreme width of the range of possible z values. On the other hand we can argue that it is more probable for the uncertainties in the basic measurements to combine in a less extreme way, some making positive contributions to δz and some negative, so that the resulting δz value is smaller than for the pessimistic assumption. This argument is valid, and we shall deal later with the question of probable uncertainty in computed quantities. For the moment, however, let us calculate that value of δz which represents the widest range of possibility for z. Such an approach, if pessimistic, is certainly safe, since, if δx, δy, etc., represent limits within which we are "almost certain" the actual values lie, then the calculated δz will give those limits within which we are equally certain that the actual value of z lies.

The most instructive initial approach uses the elementary substitution method, and we shall use this for the first two functions.

(a) Sum of Two or More Variables

Consider

$$z = x + y$$

The uncertainty in z will be obtained from

$$z_0 \pm \delta z = (x_0 \pm \delta x) + (y_0 \pm \delta y)$$

and the maximum value of δz is obtained by choosing similar signs throughout. Thus,

$$\delta z = \delta x + \delta y$$

As might be expected, the uncertainty in the sum is just the sum of the individual uncertainties. This can be expressed in terms of the relative uncertainty,

$$\frac{\delta z}{z} = \frac{\delta x + \delta y}{x + y}$$

but no increased clarification is achieved.

(b) Difference of Two Variables

Consider

$$z = x - y$$

As in the case above, δz will be obtained from

$$z_0 \pm \delta z = (x_0 \pm \delta x) - (y_0 \pm \delta y)$$

Here, however, we can obtain the maximum value of δz by choosing the *negative* sign for δy, giving, once again,

$$\delta z = \delta x + \delta y$$

We can see from this equation that, when x_0 and y_0 are close together and $x - y$ is small, the relative uncertainty can rise to very large values. This is, at best, an unsatisfactory situation, and the precision can be low enough to destroy the value of the measurement. The condition is particularly hazardous since it can arise unnoticed. It is perfectly obvious that, if it were possible to avoid it, no one would attempt to measure the length of my notebook by measuring the distance of each edge from a point a mile away and then subtracting the two lengths. However, it can happen that a desired result is to be obtained by subtraction of two measurements made separately (two thermometers, clocks, etc.), and the character of the measurement as a difference may not be strikingly obvious. Consequently, all measurements involving differences should be treated with the greatest caution. The way to avoid the difficulty, clearly, is to measure the difference directly, rather than obtaining it by subtraction between two measured quantities. For example, if one has an apparatus within which two points are at potentials above ground of $V_1 = 1500$ V and $V_2 = 1510$ V, respectively, and the required quantity is $V_2 - V_1$, only a very high quality voltmeter would permit the values of V_2 and V_1 to be measured with the exactness required to achieve even 10% precision in $V_2 - V_1$. On the other hand, an ordinary 10-V table voltmeter, connected between the two points and measuring $V_2 - V_1$ directly, will immediately give the desired result with 2% or 3% precision.

2–9 GENERAL METHOD FOR UNCERTAINTY IN FUNCTIONS OF TWO OR MORE VARIABLES

The last two examples, treated by the elementary method, suggest that, once again, the differential calculus may offer considerable simplification of the treatment. It is clear that, if we have

$$z = f(x, y)$$

the appropriate quantity for calculating δz is the total differential dz. This is given by

$$dz = \frac{\partial f}{\partial x} dx + \frac{\partial f}{\partial y} dy \qquad (2–2)$$

We shall take this differential and treat it as a finite difference δz that can be calculated from the uncertainties δx and δy. Thus,

$$\delta z = \frac{\partial f}{\partial x} \delta x + \frac{\partial f}{\partial y} \delta y$$

and the derivatives $\partial f/\partial x$ and $\partial f/\partial y$ will normally be evaluated for the values, x_0 and y_0, at which δz is required. We may find that, depending on the function f, the sign of $\partial f/\partial x$ or $\partial f/\partial y$ turns out to be negative. In this case, using our pessimistic requirement for the maximum value of δz, we should choose negative values for the appropriate δx or δy, obtaining thereby a wholly positive contribution to the sum.

(a) Product of Two or More Variables

Suppose

$$z = xy$$

To use Eq. (2–2) we need the values of $\partial z/\partial x$ and $\partial z/\partial y$. They are

$$\frac{\partial z}{\partial x} = y \qquad \text{and} \qquad \frac{\partial z}{\partial y} = x$$

Thus, the value of δz is given by

$$\delta z = y\,\delta x + x\,\delta y$$

The significance of this result is more clearly seen when it is converted to the relative uncertainty

$$\frac{\delta z}{z} = \frac{\delta x}{x} + \frac{\delta y}{y}$$

Thus, when the desired quantity is a product of two variables, its relative uncertainty is the sum of the relative uncertainties of the components.

The most general case of a compound function, very commonly found in physics, involves an algebraic product that has components raised to powers. Let

$$z = x^a y^b$$

where a and b may be positive or negative, integral or fractional. In this case the formulation is greatly simplified by taking logs of both sides before differentiating. Thus,

$$\log z = a \log x + b \log y$$

whence, differentiating implicitly,

$$\frac{dz}{z} = a\frac{dx}{x} + b\frac{dy}{y}$$

As usual, we take the differentials to be finite differences and obtain

$$\frac{\delta z}{z} = a\frac{\delta x}{x} + b\frac{\delta y}{y}$$

Note that this process gives the relative uncertainty directly, and this is frequently convenient. If the absolute uncertainty δz is required, it can be evaluated simply by multiplying the relative uncertainty by the computed value z_0, which is normally available. This form of implicit differentiation still offers the simplest procedure even when z itself is raised to some power. For, if the equation reads

$$z^2 = xy$$

it is unnecessary to rewrite it

$$z = x^{1/2}y^{1/2}$$

and work from there, because, by taking logs,

$$2 \log z = \log x + \log y$$

whence

$$2\frac{\delta z}{z} = \frac{\delta x}{x} + \frac{\delta y}{y}$$

giving $\delta z/z$ as required.

(b) Quotients

These can be treated as products in which some of the powers are negative. As before, the maximum value of δz will be obtained by neglecting negative signs in the differential and combining all the terms additively.

   If a function other than those already listed is encountered, some kind of differentiation will usually work. It is frequently convenient to differentiate an equation implicitly, thereby avoiding the requirement to calculate the unknown quantity explicitly as a function of the other variables. For example, consider the thin-lens equation

$$\frac{1}{f} = \frac{1}{o} + \frac{1}{i}$$

where the focal length f is a function of object distance o and image distance i, the measured quantities. We can differentiate the equation implicitly to obtain

$$-\frac{df}{f^2} = -\frac{do}{o^2} - \frac{di}{i^2}$$

It is now possible to calculate df/f directly and more easily than by writing f explicitly as a function of o and i and differentiating. In this way we can prepare a formula for the uncertainty into which all the unknowns can be inserted directly. Make sure that appropriate signs are used so that all contributions to the uncertainty add positively to give outer limits of possibility for the answer.

If the function is so big and complicated that we cannot obtain a value for δz in general, we can always take the measured values, x_0, y_0, etc., and work out z_0. We can then work out two different answers, one using the actual, numerical values of $x_0 + \delta x$, $y_0 + \delta y$ (or $y_0 - \delta y$ if appropriate), etc., to give one of the outer values of z and the other using $x_0 - \delta x$, etc. These two values will correspond to the limits on z, and we shall know the value of δz.

2–10 COMPENSATING ERRORS

A special situation can appear when compound variables are involved. Consider, for example, the well-known relation for the angle of minimum deviation D_m for a prism of refractive index n and vertical angle A:

$$n = \frac{\sin \frac{1}{2}(A + D_m)}{\sin \frac{1}{2}A}$$

If A and D_m are measured variables with uncertainties δA and δD_m, the quantity n will be the required answer, with an uncertainty δn. It would be fallacious, however, to calculate the uncertainty in $A + D_m$, then in $\sin \frac{1}{2}(A + D_m)$, and combine that with the uncertainty in $\sin \frac{1}{2}A$, treating the function as a quotient of two variables. This can be seen by thinking of the effect on n of an increase in A. *Both* $\sin \frac{1}{2}(A + D_m)$ and $\sin \frac{1}{2}A$ increase, and the change in n is not correspondingly large. The fallacy lies in applying the methods of the preceding sections to variables that are not independent (e.g., $A + D_m$ and A). The cure is either to reduce the equation to a form in which the variables are all independent, or else to go back to first principles and use Eq. (2–2) directly. Cases which involve compensating errors should be watched carefully, since they can, if treated incorrectly, give rise to errors in uncertainty calculations that are hard to detect.

2–11 SIGNIFICANT FIGURES

Since computations tend to produce answers consisting of long strings of numbers, we must be careful to quote the final answer sensibly. If, for example, we are given the voltage across a resistor as 15.4 ± 0.1 volts and the current as 1.7 ± 0.1 amps, we can calculate a value for the resistance. The ratio V/I comes out on my calculator as 9.0588235 ohms. Is this the answer? Clearly not. A brief calculation shows that the absolute uncertainty in the resistance is close to 0.59 ohms. So, if the first two places of decimals in the value for the resistance are uncertain, the rest are clearly meaningless. A statement like $R = 9.0588235 \pm 0.59$ ohms is, therefore, nonsense. We should quote our results in such a way that the answer and its uncertainty are consistent, e.g., $R = 9.06 \pm 0.59$ ohms.

But is even this statement really valid? Remember that the originally quoted uncertainties for V and I had the value ± 0.1, containing one significant figure. If we did now know these uncertainties any more precisely, we have no right to claim two significant figures for the uncertainty in R. Our final, valid, and self-consistent statement is, therefore, R = 9.1 $\pm$ 0.6 ohms. Only if we had real reason to believe that our original uncertainty was accurate to two significant figures could we lay claim to two significant figures in the final uncertainty and a correspondingly more precisely quoted value for R. In general terms we must make sure that our quoted values for uncertainty are consistent with the precision of the basic uncertainties and that the number of quoted figures in the final answer is consistent with the uncertainty of that final answer. We must avoid statements like $z = 1.234567 \pm 0.1$ or $z = 1.2 \pm 0.000001$.

PROBLEMS

1. I use my meter stick to measure the length of my desk. I am sure that the length is not less than 142.3 cm and not more than 142.6 cm. State this measurement as a central value $\pm$ uncertainty. What is the relative uncertainty of the measurement?

2. I read a needle-and-scale voltmeter and ammeter and assess the range of uncertainty visually. I am sure the ammeter reading lies between 1.24 and 1.25 A and the voltmeter reading between 3.2 and 3.4 V. Express each reading as a central value $\pm$ uncertainty and evaluate the relative uncertainty of each measurement.

3. My digital watch gives a time reading as 09:46. What is the absolute uncertainty of the measurement?

4. If I can read a meter stick with absolute uncertainty $\pm$ 1 mm, what is the shortest distance that I can measure if the relative uncertainty is not to exceed (a) 1%, (b) 5%?

5. I use a thermometer graduated in $\frac{1}{5}$ degree Celsius to measure outside air temperature. Measured to the nearest $\frac{1}{5}$ degree, yesterday's temperature was 22.4° Celsius and today's is 24.8° Celsius. What is the relative uncertainty in the temperature difference between yesterday and today?

6. The clock in the lab has a seconds hand that moves in one-second steps. I use it to measure a certain time interval. At the beginning of the interval it reads 09:15:22 (hours:minutes:seconds) and at the end it reads 09:18:16. What is the relative uncertainty of the measured time interval?

7. For the desk mentioned in Problem 1 I measure the width, and I am sure the measurement lies between 78.2 cm and 78.4 cm. What is the absolute uncertainty of the calculated area of the desk top?

8. In measuring the resistance of a resistor, the voltmeter reading was 15.2 $\pm$ 0.2 V and the ammeter reading was 2.6 $\pm$ 0.1 A. What is the absolute uncertainty of the resistance calculated using the equation $R = V/I$?

9. A simple pendulum is used to measure the acceleration of gravity using $T = 2\pi\sqrt{l/g}$. The period T was measured to be 1.24 ± 0.02 sec and the length to be 0.381 ± 0.002 m. What is the resulting value for g with its absolute and relative uncertainty?

10. An experiment to measure the density, d, of a cylindrical object uses the equation $d = m/\pi r^2 l$, where

$$m = \text{mass} = 0.029 \pm 0.005 \text{ kg}$$

$$r = \text{radius} = 8.2 \pm 0.1 \text{ mm}$$

$$l = \text{length} = 15.4 \pm 0.1 \text{ mm}$$

What is the absolute uncertainty of the calculated value of the density?

11. The focal length, f, of a thin lens is to be measured using the equation $1/o + 1/i = 1/f$, where

$$o = \text{object distance} = 0.154 \pm 0.002 \text{ m}$$

$$i = \text{image distance} = 0.382 \pm 0.002 \text{ m}$$

What is the calculated value for focal length, its absolute uncertainty, and its relative uncertainty?

12. A diffraction grating is used to measure the wavelength of light using the equation $d \sin \theta = \lambda$. The value of θ is measured to be $13° \, 34' \pm 2'$. Assuming that the value of d is 1420×10^{-9} m and that its uncertainty can be ignored, what are the absolute and relative uncertainties in the value of λ?

13. A value is quoted as 14.253 ± 0.1. Rewrite it with the appropriate number of significant figures. If the value is quoted as 14.253 ± 0.15, how should it be written?

14. A value is quoted as $6.74914 \pm 0.5\%$. State it as a value $\pm$ absolute uncertainty, both with the appropriate number of significant figures.

3

Statistics
of Observation

3–1 STATISTICAL UNCERTAINTY

In the preceding chapter we considered measurements in which the uncertainty could be estimated by personal judgment. In these, supposing that we have judged the situation accurately, repeated measurements should give consistent answers. Sometimes, however, repeated measurements give clearly different answers. For example, if we are using a Geiger counter and scaler to measure the activity of a radioactive source, and we decide, with given geometry, to obtain the number of counts in a 10-second interval, we would find that the results obtained by counting in successive 10-second intervals are *not* the same. We can encounter the same situation in measurements that involve visual judgment. If, for example, we wish to find the image formed by a thin lens, we may be unable to judge the position of the image accurately enough to obtain repeatedly the same reading on a good, finely divided distance scale. Whether the fluctuation is intrinsic to the system under investigation (as in the radioactive source, where the fluctuation arises from the basic nature of radioactive decay) or arises from our difficulty in making a measurement, we must find out how to make sensible statements about measurements that show such fluctuations.

What kind of statement will it be possible to make? No longer can we make such statements as we made earlier having the form "I am virtually certain that the answer lies within the interval . . . ." In fact, apart altogether from the impossibility of obtaining "right" answers, we shall find out that the difficulty lies not so much in constructing sensible answers as in knowing the sensible questions to ask. We shall discover that the only sensible questions involve, as before, intervals along our scale of values—this time, however, interpreted in terms of probabilities instead of certainty. Our search for a solution will be fairly lengthy, but at the end the answer will turn out to be simple and elegant.

To start our search, let us go back to the basic situation. Let us assume that we have made a single measurement and, in order to check our work, that we have made the measurement a second time and obtained a different answer. What are we supposed to do? We have no way of saying that one answer is "right" and the other "wrong." Which one would we choose to be "right"? In response to this ambiguity the natural reaction would be to try a third time, hoping, perhaps, that the third reading will confirm one or other of the first two. Very likely it will not be so obliging and will simply add to the confusion by supplying a third possibility. Faced with growing complexity, we could decide to keep on making measurements to see what happens. Let us suppose that our curiosity has prompted us to make a substantial number of repeated measurements, say 100, and we now ask: what is the answer? As was mentioned earlier, it is more significant to ask: what is the question? That depends very much on the use to which we wish to put the measurements. A physicist measuring the position of an optical image may be seeking something he would like to consider as the "right" answer. A person measuring the activity of a radioactive source may wish to use it in a way that requires him to know the number of counts he will obtain in a certain 10-second interval tomorrow. A sociologist counting political opinions wishes to predict the outcome of the next election, etc. There is no single question and no unique answer. The treatment we give our fluctuating numbers depends on circumstances. Let us now consider some of the possibilities.

3–2 HISTOGRAMS AND DISTRIBUTIONS

Let us assume that we have made 100 measurements of some quantity and that we must now report our results. The first response to the question, "What did you obtain?" is the rather feeble reply, "I made the measurement 100 times and here are the 100 answers." This is perhaps free of error but is hardly helpful. Our audience will find it difficult to make any sense out of a plain list of numbers, and questions will naturally arise, such as: are there any regularities in the numbers, do any appear more frequently than others, etc.? In order to show the characteristics of the measurements more clearly, some kind of graphic display would clearly be helpful.

One common mode of presentation is the histogram. To construct this diagram we divide the scale along which the measurements are spread into intervals, and we count the readings that fall within each interval. We then plot these numbers on a vertical scale against the intervals themselves. It is conventional to use a bar diagram to indicate the number of readings, and the result will be similar to Fig. 3–1. At once we improve our comprehension of the measurements enormously, because we can see at a glance how the values are *distributed* along the scale. This distribution is the key to satisfactory interpretation of the measurements. Usually we find that the readings tend to occur more frequently in the middle of the range, and, if this is so and we are unable to make any other sensible statement, we can always content ourselves with the simple assertion that the observations have "central tendency." This may suffice, and when we have drawn the histogram we may be able to stop. Many

Table 1

85	109	114	121	127	131
92	109	114	121	127	132
96	110	114	122	127	133
97	110	115	122	127	134
97	111	116	122	128	134
97	111	116	122	128	134
100	111	116	122	128	134
101	111	117	123	128	135
101	111	117	123	128	136
102	112	118	123	128	137
102	112	118	123	130	137
103	112	119	123	130	137
103	113	119	124	130	144
105	113	120	124	130	148
106	113	120	124	130	149
106	113	120	125	130	
107	113	120	125	131	
108	113	121	125	131	
108	114	121	126	131	

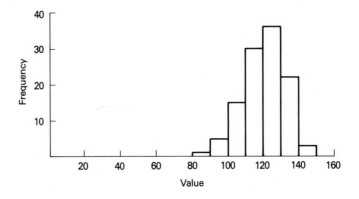

Figure 3-1   A set of observations and its histogram.

results from measurement processes are presented simply by offering the histogram; the reader can view the distribution and draw his own conclusions.

3–3 CENTRAL VALUES OF DISTRIBUTIONS

Frequently, however, we wish to go further and, as a substitute for the whole histogram, find some shorthand way of describing the distribution without actually showing the whole diagram. We can seek answers, therefore, to questions such as: what *single* number best characterizes the complete group of observations? There are several candidates for such designation, and we choose one on the basis of the future use of the information. The various possibilities are:

(a) Mode

Most distributions have a peak near the center. If this peak is well defined, the value on the horizontal scale at which it occurs is called the **mode** of the distribution. Whenever we wish to draw attention to such central concentration in our measured values, we quote the modal value. Sometimes a distribution will show two peaks; we call it a **bimodal** distribution and quote the two modal values.

(b) Median

If we place all our readings in numerical order and divide them in the middle into two equal parts, the value at which the dividing line comes is called the **median.** Since it is obvious that areas under distribution graphs represent numbers of observations (the left-hand bar in Fig. 3–1 represents 5 observations, the second from the left represents 9, so that the two together represent 14, and so on), the median is that value at which a vertical line divides the distribution into two parts of equal area. The median is frequently quoted in sociological work; people talk about median salaries for certain groups of employees, etc.

(c) Mean

The third of the commonly quoted numbers is the familiar arithmetic average or mean. For a group of N observations, x_i, the mean $\bar{x}$ is defined by

$$\bar{x} = \frac{\sum x_i}{N} \tag{3–1}$$

We shall discover that, for our purposes, the mean is the most useful of the three quantities we have defined.

   Notice that, for a symmetrical distribution, the mean, median, and mode all coincide at the center of the distribution. If, on the other hand, the distribution is not symmetrical, each will have a separate value. For the histogram shown in Fig. 3–1, the values of the mean, median, and mode are shown in Fig. 3–2, which illustrates their relationship to the distribution. If the distribution is markedly asymmetric, the difference between the mode, median, and mean can be substantial. Consider, for example, the distribution of family income in a country. The presence of the millionaires, although few in number, has an effect on the mean that counterbalances many members of the popula-

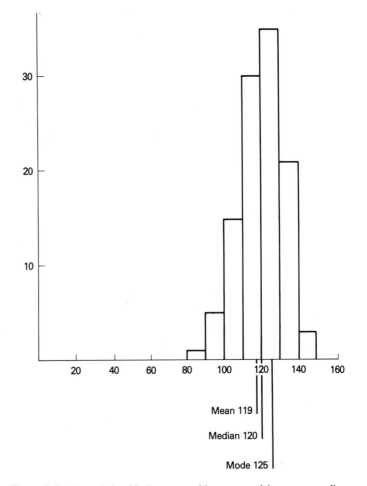

Figure 3-2 The relationship between a histogram and its mean, median, and mode.

tion at the low end of the salary scale. The mode and the mean thus differ substantially. This example illustrates the care required in interpreting quoted statistics; people who quote statistics frequently do so in the way that best suits their particular purpose.

3–4 THE BREADTH OF DISTRIBUTIONS

Let us now turn to the question: to what extent is our chosen number representative of the distribution as a whole? That is, how reliable is it to use a single number to substitute for a whole distribution? At the present stage we have no justification to offer for the procedures that will be described. We shall rely, instead, on an intuitive feeling that, the broader is the distribution, the less significance we can ascribe to any one of the three central values. On the other hand, the narrower the distribution, the more we feel entitled to confidence in the mean, mode, or median as significant quantities for the distribution.

Let us, therefore, construct a quantity that will be a measure of the breadth of the distribution. We could invent many such quantities, but, for reasons that need not concern us at the moment, we shall define a quantity that is almost universally used. We define the **standard deviation** of the distribution, S, to be

$$S = \sqrt{\frac{\Sigma \, (\bar{x} - x_i)^2}{N}} \tag{3–2}$$

The definition is to some extent arbitrary, for, in defining a measure of the breadth of the distribution, we could have chosen other powers for the quantity $(\bar{x} - x_i)$, and we could have chosen other denominators. There are, however, reasons for these choices; these reasons and the significance of the standard deviation will become clear shortly.

We can pause at this stage to summarize the progress so far. If we have made repeated measurements of a quantity and wish to state the result in numerical terms, we can do a number of things: (a) we can show the histogram, (b) we can quote the mode, median, or mean as a measure of the location of the distribution, and (c) we can quote the standard deviation as a measure of the confidence we can place in the results. We sometimes leave the outcome of a measuring process in this form; the quantities involved are universally understood, and the procedure is acceptable.

For our present purpose, however, we seek more detailed, numerical interpretation of the quoted values.

3–5 SIGNIFICANCE OF THE MEAN AND STANDARD DEVIATION

In this and the following sections we shall, for reasons that will become clear, ignore the mode and median and restrict ourselves to numerical interpretation of the mean and the standard deviation. Since the presence of random fluctuation has denied us the opportunity to identify a realistic interval within which we can feel certain our answer lies, we must alter our expectations of the measuring process. As we have said before, it is not so much a matter of obtaining sensible answers to questions as of knowing the sensible questions to ask. Specifically, of course, it is not sensible to ask: what is the right answer? It is not even sensible to ask: having made one hundred observations of a quantity, what shall I obtain when I make the measurement the next time? The only sensible questions involve not certainty but probability, and several different questions about probabilities are possible.

We could ask, for example: what is the probability that the 101st reading will fall within a certain range on our scale of values? That is a sensible question, and sensible answers can easily be imagined. If, for example, of our 100 original readings, a certain fraction of the values fell within that particular range, we might feel justified in choosing that fraction as the probability we seek. This would not be an unrealistic guess, and we could attempt a standardized description of our distribution by quoting the fraction of the total number of readings that fall within a specified interval, such as $x \pm S$. This would satisfactorily convey information about our set of readings to other people, but a major problem appears when we discover that our answers for probabilities are specific to our particular histogram. If we were to make another series of 100 readings, holding all the conditions the same as they were before in the hope of obtaining the same histogram, we would be disappointed. The new histogram would not duplicate the first exactly. It might have similar general characteristics with respect to location and breadth, but its detailed structure would not be the same as before, and we would obtain different answers to questions about probabilities.

How, then, are we going to find answers to our questions that have some kind of widely understood numerical significance? One solution is to abandon the attempt to describe our particular histogram and to start talking about defined, theoretical distributions. These may have the disadvantage of uncertain relevance to our particular set of observations, but there is the enormous advantage that, since they are defined, theoretical constructs, they have properties that are definite, constant, and widely understood. Many such theoretical distributions have been constructed for special purposes, but we shall restrict ourselves to one only, the Gaussian or "normal" distribution.

We use the Gaussian distribution to interpret many kinds of physical measurement, partly because the mechanical circumstances of many physical measurements are in close correspondence with the theoretical foundations of the Gaussian distribution, and partly because experience has shown that Gaussian statistics do provide a reasonably accurate description of many real events. For only one common type of physical measurement is another distribution more appropriate; in counting events like radioactive decay we must use a distribution called the Poisson distribution, but, even for it, the difference from Gaussian statistics becomes significant only at low counting rates. Further information about Poisson statistics will be found in books describing experimental methods in nuclear or high-energy physics. Apart from these special cases, we can feel relatively confident that Gaussian statistics can be usefully applied to most real measurements. We should, however, always remember that, unless we actually test our measurements for correspondence with the Gaussian distribution, we are making an assumption that Gaussian statistics are applicable, and we should remain alert to any evidence that the assumption may be invalid.

3–6 GAUSSIAN DISTRIBUTION AND SAMPLING

Even if, to use it successfully, we need not know very much about the origins of the Gaussian distribution, it is interesting to know why its derivation makes it particularly relevant to many physical measurements. The Gaussian distribution can be derived from the assumption that the total deviation of a measured quantity, x, from a central value, X, is the consequence of a large number of small fluctuations that are of random occurrence. To construct a simple model of such a situation, let us suppose that there are m such contributions to the total deviation, each of equal magnitude a and equally likely to be positive or negative. If we repeat the measuring process many times, we shall obtain a set of values that will range from $X + ma$, for a reading in which all the fluctuations happened to be positive simultaneously, to $X - ma$, if the same happened in the negative direction. For such random summation of positive and negative quantities (as in the "random walk"), we can prove that the most probable sum is zero, meaning that the most common values of x are in the vicinity of X. The distribution curve, therefore, has a peak in the middle, is symmetrical, and declines smoothly to zero at $x = X + ma$ and $x = X - ma$. If this concept is taken to the limiting case in which an infinite number of infinitesimal deviations contribute to the total deviation, the curve has the form shown in Fig. 3–3. Treating the curve solely from the mathematical point of view for the moment, its equation can be written

$$y = Ce^{-h^2(x-X)^2} \tag{3-3}$$

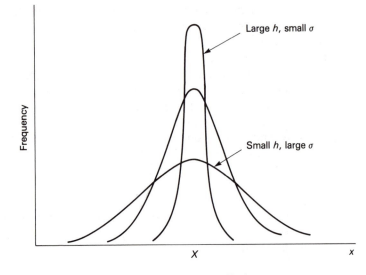

Figure 3-3   The Gaussian distribution curve.

Here the constant C is a measure of the height of the curve, since $y = C$ for $x = X$ at the center of the distribution. The curve is symmetrical about $x = X$ and approaches zero asymptotically. The quantity h obviously governs the width of the curve, since it is only a multiplier on the x scale. If h is large, the curve is narrow, and high in relation to its width; if small, the curve is low and broad. The quantity h clearly must be connected with the standard deviation, σ, of the distribution, and it can be shown that the relationship is

$$\sigma = \frac{1}{\sqrt{2}\, h} \qquad (3\text{--}4)$$

(We shall use Latin letters, e.g., S for standard deviation, for quantities associated with finite sets of actual observations, and Greek letters, such as σ, when referring to defined distributions or, as described in Sec. 3–7, to a "universe" of observations.)

Now that we have a definite equation for the distribution, all the original ambiguity about interpreting the standard deviation in terms of probability disappears, and we have definite, unique, and permanent values. For example, the area enclosed within the interval $X \pm \sigma$ for a Gaussian distribution is 68% and within the interval $X \pm 2\sigma$ it is 95%, and this is so for *all* Gaussian distributions. The relation between the σ values and areas on the distribution curve is shown in Fig. 3–4 by the lines drawn vertically at intervals of 1σ and 2σ from the central value. It is very comforting to have such definite numbers, be-

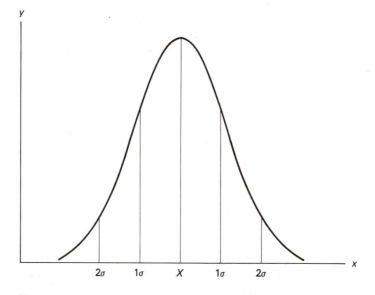

Figure 3-4   The relationship of 1σ and 2σ limits to the Gaussian distribution.

cause we can say definitely that any particular value in a Gaussian set has a 68% chance of falling within the interval $X \pm \sigma$ and a 95% chance of falling within $X \pm 2\sigma$. A more extensive account of the mathematical properties of the Gaussian distribution will be found in Appendix 1.

3–7 RELATION BETWEEN GAUSSIAN DISTRIBUTIONS AND REAL OBSERVATIONS

The results given in the preceding section provide useful, precise methods for interpreting means and standard deviations, but a problem arises when we start applying such thoughts to real measurements. Numbers like 68% and 95% refer to a theoretical construct, the Gaussian distribution, and all we have is one, or at most a few, real measurements of our desired quantity. We have, at first, no way of knowing which Gaussian distribution, with attached values of X and σ, is appropriate to our observations. So what are we to do? The answer lies in a concept which provides a bridge between the world of theoretical constructs and the world of real measurements. We invent, for a piece of apparatus or a measuring process, the concept of the infinite set of measurements which *could* be made with it. Of course, for rather obvious reasons, this infinite set of measurements will never be made, but the concept enables us to interpret our real measurements. The construct is called the "universe" or "population" for that particular measurement. Once we have made, say, 100 measurements with a particular apparatus, we have a tendency to feel that nothing exists but our 100 values. We must now invert our thinking and view our set of measurements as a "sample" of the infinitely large universe or population of measurements that could be made. The universe, however, is permanently inaccessible to us; we shall never know the universe distribution or its mean or its standard deviation. Our task will be to construct inferences about these quantities from the definitely known properties of our sample.

We shall do this on the basis of some assumptions. First, we shall assume that the universe distribution is Gaussian, and we shall call the universe mean X and the universe standard deviation σ. This assumption enables us to make statements such as: if we make just one measurement with our equipment, that one measurement has a 68% chance of falling within $X \pm \sigma$ and a 95% chance of falling within $X \pm 2\sigma$. This seems like an encouragingly exact and explicit statement, but it suffers from an overwhelming defect; we do not (and never shall) know the values of X and σ. In other words, having made only one observation of a quantity that is subject to random fluctuation, we have gained practically nothing. We can say only that our value has a 68% chance of falling within something of somewhere, which is not too helpful. Our only hope lies in obtaining some information, even if uncertain, about the universe distribution. As we have already mentioned, we are never going to be able to determine the universe distribution exactly, because that would require an infinite number of readings. We can only hope that, if we repeat our measuring process to obtain a sample from the universe, that sample will enable us to make some estimate of the universe parameters.

Since we are making the basic assumption that the universe distribution is a mathematical, defined function (whether Gaussian or some other, equally well-defined distribution), we can evaluate mathematically the properties of samples with respect to those of the universe of single observations. We shall simply state these properties of samples without proof. The reader who is curious about the mathematical derivation of these results is encouraged to turn to the standard texts on statistics, in which sections dealing with sampling theory will be found.

The properties of samples become clear if we consider the concept of repeated sampling. Consider that, with a certain piece of apparatus, we make 100 observations. This will be our first sample; let us calculate its mean and standard deviation and record them. Now let us make another set of 100 observations and record for it the mean and standard deviation. Let us continue such repetition until we have an infinite number of samples, each with its own mean and standard deviation, and let us then plot the distribution curves of the sample means and of the sample standard deviations. Of course we shall never carry out a process like this with actual observations but, knowing the mathematical function for our original universe of single readings, we can simulate such repeated sampling mathematically, and so derive the properties of the samples in comparison with those of the original universe of single readings. The results of such calculations of the distribution of sample means and sample standard deviations are shown in Fig. 3–5 and Fig. 3–6 and they will be described in the following sections.

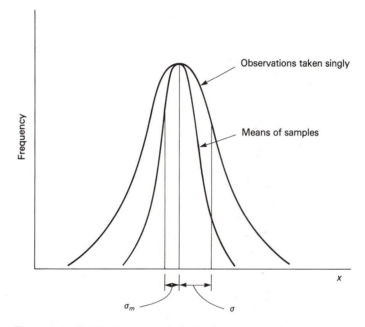

Figure 3-5   Distribution curve of single observations and sample means. (Note that the vertical scale for the two curves is not the same. They have been plotted with a common peak value solely for purposes of illustration.)

3–8 SAMPLE MEANS AND STANDARD DEVIATION OF THE MEAN

If the universe destribution of single readings is Gaussian, the theory of sampling shows that the distribution of sample means is also Gaussian. In addition, the distribution of sample means has two other very important properties. First,

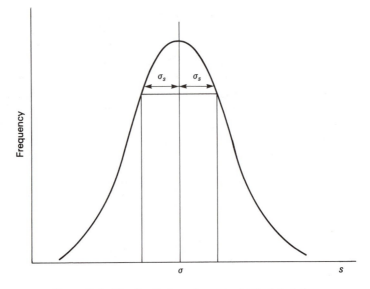

Figure 3-6   The distribution of sample standard deviations.

it is centered on X, the center of the original distribution of single readings; second, it is narrower than the original distribution. This narrowness is very significant, because it demonstrates immediately the improvement in precision that comes from samples as opposed to single readings; the means of samples cluster more closely around the universe mean than do single readings. The reduced scatter of sample means is represented by a very important quantity, the standard deviation of the distribution of sample means. This quantity is called the **standard deviation of the mean** and its value is

$$\sigma_m = \frac{\sigma}{\sqrt{N}} \qquad (3\text{-}5)$$

where N is the number of readings in the sample. Thus, a particular sample mean has a 68% chance of falling within the interval $X \pm \sigma_m$ and a 95% chance for the interval $X \pm 2\sigma_m$. These intervals are smaller than the corresponding intervals for single readings, and they supply a numerical measure of the improved precision that is available from sampling.

Note that the statement about sample means, although precise, still does not help us very much, because it still involves the unknown quantities X and σ. The resolution of this difficulty and the significance of the standard deviation of the mean will become clear very soon. In the meantime let us turn our attention briefly to the other important property of samples, the distribution of sample standard deviations.

3–9 SAMPLE STANDARD DEVIATION

The sample standard deviations also fall on a Gaussian distribution, the center of which is the universe standard deviation, σ. The distribution is illustrated in Fig. 3–6. As will become clear in a moment, however, the variance of the sample standard deviations will not concern us as much as the variance of sample means, and we shall postpone to Sec. 3–11 further discussion of the variance of sample standard deviations.

3–10 APPLICATION OF SAMPLING THEORY TO REAL MEASUREMENTS

The sample properties which we have just presented are very interesting, but how do they help us when we do not have access to the actual distributions, either for sample means or sample standard deviations? We have our lone sample with *its* mean and standard deviation, and no idea how they relate to the universe values. Our problem, therefore, is to find a connection between the theoretical results and the sample properties that allows us to infer the universe properties from the sample values. We cannot expect, obviously, to obtain exact information. In addition, we must make one basic, obviously imprecise, assumption. We assume that our single value, the sample standard deviation, provides us with a value for the universe standard deviation. In fact it can be proved that the "best estimate" of the universe standard deviation is given by the quantity

$$S = \sqrt{\frac{\Sigma\, (\bar{x} - x_i)^2}{N - 1}} \qquad (3\text{–}6)$$

This quantity is only slightly different from our original value for the standard deviation of a set of observations. The N in the denominator of the original expression has been replaced by $N - 1$, and the difference between the two quantities, obviously, is significant only for small values of N. In the future, when we talk about a sample standard deviation, we shall assume that we are using the equation in the new form and that we are really talking about the "best estimate" of the universe value σ.

Accepting our sample standard deviation as the best estimate of σ, we are now able to make a definite statement about our single sample. We can rephrase Eq. (3–5) and define

$$S_m = \frac{S}{\sqrt{N}} \qquad (3\text{–}7)$$

as our standard deviation of the mean, now a known quantity obtained from our real sample. We can now say: our sample mean $\bar{x}$ has a 68% chance of falling within $X \pm S_m$ and a 95% chance of falling within $X \pm 2S_m$. This is a statement which is close to what we want, but it is not yet completely satisfactory. It tells us something about a quantity that we know, $\bar{x}$, in terms of a quantity that we do not know, X. We really want the statement to be the other way around; we want to be able to make an assertion about our unknown, X, in terms of a quantity, $\bar{x}$, of which we do know the value. Fortunately, it is possible to prove that the above statement about probabilities can be inverted to yield our desired result. We obtain thereby the statement toward which we have been working ever since we started our discussion of the statistics of fluctuating quantities. Our final statement is: there is a 68% chance that the universe mean, X, falls within the interval $\bar{x} \pm S_m$ and a 95% chance that it falls within the interval $\bar{x} \pm 2S_m$. This is now, finally, a statement about the unknown quantity, X, in terms of wholly known quantities, $\bar{x}$ and S_m. Along our scale of x values we now have a real and known interval between $\bar{x} - S_m$ and $\bar{x} + S_m$, and we know that there is a 68% chance that our desired quantity X lies within this interval.

This statement provides us with the answer we have been seeking and brings us as close as we can come to exact information about the unperturbed value of our measured quantity. It is worth becoming familiar with the arguments that have been given in the preceding sections; there is more to measurement than simply making a few measurements and "taking the average" just because it seems to be the right thing to do.

3–11 EFFECT OF SAMPLE SIZE

In any sampling process, clearly, the larger the sample, the more precise will be our final statements. Even though the precision of a mean value increases only as the square root of the number of observations in the sample [Eq. (3–5)], it does increase, and larger samples have more precise means. There may, however, be limitations of time or opportunity, and we cannot always obtain samples of the size we would like. Usually a compromise must be sought between the conflicting demands of precision and time, and good experiment design will incorporate this compromise into the preliminary planning. Nevertheless, it may occasionally be necessary to content ourselves with small samples. In this undesirable eventuality we should be aware of the magnitude of the resulting loss of precision. There is, first, the influence on the value of the standard deviation of the mean; the smaller N is, the larger will be the value of S_m and the longer the interval on the x scale that has the 68% chance of containing the universe value X. Second, we must, for small samples, place declining faith in our use of the sample standard deviation S as the best estimate of the universe value σ. To illustrate this, recall the distribution curve for sample standard deviations that was shown in Fig. 3–6. It is worth asking:

given the existence of this distribution, how good is our "best estimate" of the universe standard deviation, and how does it vary with sample size? The answer must be based on the width of the distribution of sample standard deviations, and so we must calculate the standard deviation of this distribution. It is called the **standard deviation of the standard deviation.** (This process could obviously go on indefinitely but, thankfully, we shall stop at this stage.) The value of the standard deviation of the standard deviation, calculated mathematically from the equation of the Gaussian distribution, is

$$\sigma_S = \frac{\sigma}{\sqrt{2(N-1}} \tag{3–7}$$

The breadth of the distribution of sample standard deviations is thus related to its central value σ by the numerical factor $1/\sqrt{2(N-1)}$. As one might expect, therefore, the accuracy of our sample standard deviation as the best estimate of the universe value is dependent on the sample size. For example, with a sample size of 10, Eq. (3–8) shows that our S value from the sample has a 68% chance of falling within a range of $\pm\sigma/\sqrt{18}$, approximately $\pm\sigma/4$, about the universe value σ. Correspondingly, the interval that has a 95% chance of containing our sample mean is as wide as $\sigma/2$ about the universe value σ. This does not represent high precision of measurement. We have, therefore, confirmation of the warning given earlier: statistical exercises with small samples should be undertaken only when no alternative exists. In order to provide an overall feeling for the reliability of σ estimates from samples of differing size, Table 3–1 contains some typical values of $\sqrt{2(N-1)}$ for various values of N.

TABLE 3–1   Accuracy of σ Estimates
from Samples of Varying Size

68% Confidence		95% Confidence	
N	$\sqrt{2(N-1)}$	N	$\sqrt{2(N-1)}$
2	1.4	2	0.7
3	2.0	3	1.0
4	2.4	4	1.2
5	2.8	5	1.4
6	3.1	6	1.6
7	3.4	7	1.7
8	3.7	8	1.8
9	4.0	9	2.0
10	4.2	10	2.1
15	5.2	15	2.6
20	6.1	20	3.2
50	9.8	50	4.9
100	14.1	100	7.0

These values are illustrated in Fig. 3–7 for $N = 3$, $N = 10$, and $N = 100$. The $\pm 1\sigma/S$ limits are marked on these curves, showing, for various sample sizes, the intervals within which there is a 68% probability that our single sample standard deviation lies. For values of N less than about 10, it is clear that the intervals for 68% or 95% probability become so large in comparison with the central value that it is almost pointless to attempt an estimate of σ. It is, therefore, rarely worth attempting any kind of statistical analysis with samples containing fewer than about 10 observations. In any case, when reporting the outcome of statistical work, it is essential to quote the sample size. If we intend our values for the mean and standard deviation of the mean to be interpreted in accordance with the 68% and 95% prescription, we must give our reader the opportunity to judge the accuracy of our estimates.

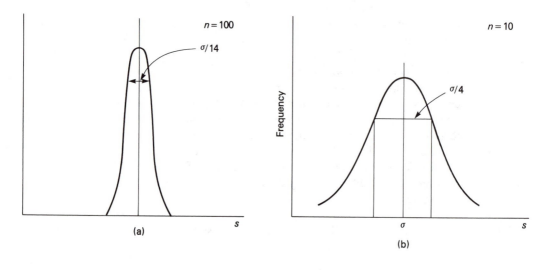

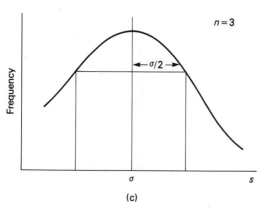

Figure 3-7 Sample standard deviation distributions for samples of various sizes.

3–12 STANDARD DEVIATION OF COMPUTED VALUES

In Chap. 2 we considered the uncertainty of computed values z, and we assumed that the uncertainty of the basic measurements constituted intervals within which we were almost certain that the values lay. We calculated the maximum range of variability of the computed answer on the pessimistic assumption that the errors in the various measured values combined, in a "worst-case" fashion, to drive the computed answer as far away from the central value as it can go. We have already suggested that this represents an unrealistically pessimistic approach and that a more useful quantity would be a "probable" value for δz based on the various probabilities associated with deviation of the basic quantities x, y, etc. from their central values. The limits given by this quantity will, naturally, be smaller than $\pm \delta z$, but we hope to find actual, numerical significance for them. Such statistical validity will be available only if the uncertainties in x and y have statistical significance, and we shall assume in the following calculations that the measurements of x and y have been sufficiently numerous to justify a calculation of the standard deviaions S_x and S_y. We hope now to calculate a value for S_z that will have the same significance for z values as S_x and S_y had for x and y.

We must, however, first inquire what we mean by S_z. We assume that the measurement has taken the form of pairs of observations x, y that were obtained by repetition of the observing process under identical conditions (for example, the current through and the potential across a resistor, measured for the purpose of calculating the resistance R). Each pair of observations will provide a value of z, and, if repetition yielded N pairs, we shall have a set of N values of z that are distributed in accordance with the fluctuations in the basic measurements. The quantity we require, S_z, is the standard deviation of this set of z values. These individual values of z may never be calculated individually, because a simpler mode of calculation exists. We can calculate the means $\bar{x}$ and $\bar{y}$, of the sets of x and y values and obtain $\bar{z}$ directly using the assumption (valid if S_x, S_y, and S_z are small compared, respectively, with $\bar{x}$, $\bar{y}$, and $\bar{z}$) that

$$\bar{z} = f(\bar{x}, \bar{y})$$

Nevertheless, that distribution of z values provides the significance of the S_z that we are about to calculate.

If we assume that the universes of the x, y, and z values have a Gaussian distribution, the quantity σ_z (of which we are about to calculate the best estimate in terms of various S values) will have the usual significance; i.e., any z value will have a 68% chance of falling within $\pm \sigma_z$ of the central value. As before, let

$$z = f(x, y)$$

and consider perturbations δx and δy which lead to a perturbation δz in the computed z value. The value of δz will be given by

$$\delta z = \frac{\delta z}{\delta x} \, \delta x + \frac{\delta z}{\delta y} \, \delta y$$

This perturbation can be used to calculate a standard deviation for the N different z values, since

$$S_z = \sqrt{\frac{\Sigma \, (\delta z)^2}{N}}$$

Thus

$$S_z^2 = \frac{1}{N} \Sigma \left(\frac{\partial z}{\partial x} \, \delta x + \frac{\partial z}{\partial y} \, \delta y \right)^2$$

$$= \frac{1}{N} \Sigma \left(\left(\frac{\partial z}{\partial x}\right)^2 (\delta x)^2 + \left(\frac{\partial z}{\partial y}\right)^2 (\delta y)^2 + 2 \frac{\partial z}{\partial x} \frac{\partial z}{\partial y} \, \delta x \, \delta y \right)$$

$$= \left(\frac{\partial z}{\partial x}\right)^2 \frac{1}{N} \Sigma \, (\delta x)^2 + \left(\frac{\partial z}{\partial y}\right)^2 \frac{1}{N} \Sigma \, (\delta y)^2 + \frac{2}{N} \frac{\partial z}{\partial x} \frac{\partial z}{\partial y} \, \delta x \, \delta y$$

But

$$\frac{1}{N} \Sigma \, (\delta x)^2 = S_x^2 \quad \text{and} \quad \frac{1}{N} \Sigma \, (\delta y)^2 = S_y^2$$

and, since δx and δy may be considered for the present purpose to be independent perturbations,

$$\Sigma \, (\delta x \, \delta y) = 0$$

Thus, finally,

$$S_z = \sqrt{\left(\frac{\partial z}{\partial x}\right)^2 S_x^2 + \left(\frac{\partial z}{\partial y}\right)^2 S_y^2} \tag{3-8}$$

If z is a function of more than two variables, the equation is extended by adding similar terms. Thus, if the components of a calculation have standard deviations with some degree of reliability, a value can be found for the probable uncertainty of the answer, where "probable" has real numerical significance.

The calculation has been performed in terms of the variance or standard deviation of the x and y distributions. In actual practice, however, we do not use the sample variance directly; we must calculate the best estimates of σ_x, σ_y, etc., and, in accordance with Eq. (3–6), we would use the modified value for standard deviation with denominator $N - 1$ instead of N. The final result would then be a best estimate for σ_z. The standard deviation of the mean for z can then be calculated by direct use of Eq. (3–5) and will give the limits that have a 68% chance of containing the unperturbed value.

Note that most experiments are not carried out in accordance with the restricted assumptions of the above development. If, for example, we are studying the flow rate of water through a pipe, we would measure the flow rate, pipe radius, and pipe length independently and would choose the number of readings in each sample on the basis of the intrinsic precision of the measurement. We cannot, therefore, use Eq. (3–8) directly, since the various S's are not compatible. The solution is to calculate the standard deviation of the mean for each of the elementary quantities first. If these are used in Eq. (3–8), the result of the calculation will be immediately a standard deviation of the mean for z.

3–13 STANDARD DEVIATION OF COMPUTED VALUES: SPECIAL CASES

Let us now apply Eq. (3–8) to a few common examples. In all the following cases the various S's are all assumed to be best estimates of the appropriate universe value σ.

(a) Sum of Two Variables

$$z = x + y$$

Here

$$\frac{\partial z}{\partial x} = 1, \qquad \frac{\partial z}{\partial y} = 1$$

and

$$S_z = \sqrt{S_x^2 + S_y^2}$$

Note that this result provides justification for Eq. (3–5). The mean value for the sample, $\Sigma \, (x_i/N)$, is just such a function as $z = x + y$, where x and y happen to be independent measurements of the same quantity. Thus, if

$$z = \frac{1}{N}(x_1 + x_2 + x_3 + \cdots)$$

$$\frac{\partial z}{\partial x_1} = 1/N \qquad \frac{\partial z}{\partial x_2} = 1/N \qquad \text{etc.}$$

and

$$S_z = \sqrt{\left(\frac{1}{N}\right)^2 S_x^2 + \left(\frac{1}{N}\right)^2 S_x^2 + \cdots}$$
$$= \sqrt{NS_x^2/N^2} = S_x/\sqrt{N}$$

(b) Difference of Two Variables

$$z = x - y$$

Here

$$\frac{\partial z}{\partial x} = 1, \qquad \frac{\partial z}{\partial y} = -1$$

but, again,

$$S_z = \sqrt{S_x^2 + S_y^2}$$

Recalling Sec. 2–8(b), we note that the earlier discussion of measured differences is still valid.

(c) Product of Two Variables

$$z = xy$$

Here

$$\frac{\partial z}{\partial x} = y, \qquad \frac{\partial z}{\partial y} = x$$

and

$$S_z = \sqrt{y^2 S_x^2 + x^2 S_y^2}$$

The specific value of S_z at the particular values, x_0 and y_0, of x and y can be obtained by substituting x_0 and y_0 in the equations. As was the case for uncertainty in products, the equation is more clearly expressed in terms of relative values of S. We obtain

$$\frac{S_z}{z} = \sqrt{\frac{S_x^2}{x^2} + \frac{S_y^2}{y^2}}$$

(d) Variables Raised to Powers

Here

$$z = x^a$$

$$\frac{\partial z}{\partial x} = ax^{a-1}$$

and

$$S_z = \sqrt{a^2 x^{2(a-1)} S_x^2}$$

Again this is more instructive when expressed in terms of relative values:

$$\frac{S_z}{z} = \sqrt{\frac{a^2 S_x^2}{x^2}}$$

$$= a\frac{S_x}{x}$$

(e) The General Case of Powers and Products

$$z = x^a y^b$$

The results of the two preceding sections can obviously be extended to give the result

$$\frac{S_z}{z} = \sqrt{\left(\frac{aS_x}{x}\right)^2 + \left(\frac{bS_y}{y}\right)^2}$$

In contrast to the case of combined uncertainty, negative powers in the original function need not be given special consideration; in the equation for S_z powers occur in squared form and automatically make a positive contribution.

If a function other than those listed above is encountered, the use of Eq. (3–8) will yield the desired result. Incidentally, we may note that, for a function of a single variable, Eq. (3–8) reduces to the same form as for uncertainties, Eq. (2–1). This correspondence is predictable for a situation in which we do not have the probability-based interplay between two or more variables.

Finally, although we listed in Sec. 2–5 to Sec. 2–9 a number of different approaches to the calculation of outer limits for uncertainties, the standard deviation of z is a uniquely defined quantity, and there is no alternative to the use of Eq. (3–8).

3–14 COMBINATION OF DIFFERENT TYPES OF UNCERTAINTY

Unfortunately for the mathematical elegance of the development, we frequently require the uncertainty in a computed result which contains quantities having different types of uncertainty. We may require the uncertainty in

$$z = f(x, y)$$

where, for example, x is a quantity to which have been assigned outer limits, $\pm \delta x$, within which we are "almost certain" that the actual value lies, while y is a quantity whose uncertainty is statistical in nature, a sample standard deviation, S_y, perhaps, or a standard deviation of the mean, $S_y/\sqrt{N}$. We require an uncertainty for z. Our initial difficulty is even to define the uncertainty in z. We are trying to combine two quantities which have, in effect, completely different distribution curves. One is the standard Gaussian function; the other is a rectangle, bounded by the values $x_0 + \delta x$ and $x_0 - \delta x$ and flat on top, because the actual value of x is equally likely to be anywhere within the interval $x_0 \pm \delta x$. Any general method of solving this problem is likely to be far too complex for general use, but a simple approximation is available using the following procedure.

In the calculation for z we use the sample mean, $\bar{y}$, for the y value, implying that the universe mean has an approximately $\frac{2}{3}$ chance of falling within the interval, $\bar{y} \pm S_y/\sqrt{N}$. Let us, therefore, calculate limits for x that also have a $\frac{2}{3}$ probability of enclosing the actual value. Since the probability distribution for x is rectangular, $\frac{2}{3}$ of the area under the distribution curve is enclosed by limits that are separated by a distance equal to $\frac{2}{3}$ of the total range of possibility, i.e., $\frac{2}{3}$ of 2 δx. The total width of the region for $\frac{2}{3}$ probability is, therefore, $\frac{4}{3} \delta x$ and the uncertainty limits are $\pm \frac{2}{3} \delta x$.

The quantity $\frac{2}{3} \delta x$ is, therefore, compatible with $S_y/\sqrt{N}$, since both refer to $\frac{2}{3}$ probability. Equation (3–9) can now be used, inserting $\frac{2}{3} \delta x$ for the value of the standard deviation of the mean for x and $S_y/\sqrt{N}$ for the y function. This will yield a value for uncertainty in z which can be interpreted in accordance with the $\frac{2}{3}$ prescription. Note, however, that the limits for 95% probability are not simply twice as wide as those for $\frac{2}{3}$ probability; they would have to be calculated separately using the above method.

3–15 REJECTION OF READINGS

One last, practical property of distribution curves concerns outlying values. There is always the possibility of making an actual mistake, perhaps in misreading a scale or in accidentally moving an instrument between setting and reading. We encounter the temptation, therefore, to assign some such cause to a single reading that is well separated from an otherwise compact group of values. This is, however, a dangerous temptation, since the Gaussian curve does permit values remote from the central part of the curve. Furthermore, once we admit the possibility of pruning the observations, it can become very difficult to know where to stop. We are dependent, therefore, on the judgment of the experimenter. This is not unreasonable, since the experimenter knows more about the measurement than anyone else, but criteria for making the choices can be helpful. Many empirical "rules" for rejection of observations have been formulated, but they must be used with discretion. It would be foolish to use a rule to reject one reading which was just outside the limit set by the rule if there are other readings just inside it. There is also the possibility that extra information relating to the isolated reading was noted at the time it was made, and this can help us decide in favor of retention or rejection.

The guidance we desire in making such decisions can be found in the properties of the Gaussian distribution. In a Gaussian distribution the probability of obtaining readings outside the 2σ limits is 5% (as we have seen before), outside 3σ limits it is $\frac{1}{3}$%, and outside 4σ limits the chance is no more than 6×10^{-5}. The decision to reject is still the responsibility of the experimenter, of course, but we can say, in general terms, that readings falling outside 3σ limits are likely to be mistakes and candidates for rejection. However, a problem can arise because of our lack of information about the universe of readings and its parameters X and σ. The better our knowledge of σ, the more confident we can be that any far-out and isolated reading arises from a genuinely extraneous cause such as personal error, malfunction of apparatus, etc. Thus, if we make 50 observations that cluster within 1% of the central value and then obtain one reading that lies at a separation of 10%, we can be fairly safe in suggesting that this last reading did not belong to the same universe as the preceding 50. The basic requirement, before any rejection is justified, is confidence in the main distribution of readings. Clearly, there is no justification for taking two readings and then rejecting a third measurement on the basis of a 3σ criterion. Unless the case for rejection is completely convincing, the best course is to retain all readings, whether we like them or not.

It is wise also to remember that many of the greatest discoveries in physics had their origin in outlying measurements.

PROBLEMS

The following observations of angles (in minutes of arc) were made while measuring the thickness of a liquid helium film. Assume that the observations show random uncertainty, that they are a sample from a Gaussian universe, and use them in Problems 1–14.

34	35	45	40	46
38	47	36	38	34
33	36	43	43	37
38	32	38	40	33
38	40	48	39	32
36	40	40	36	34

1. Draw the histogram of the observations.
2. Identify the mode and the median.
3. Calculate the mean.
4. Calculate the best estimate of the universe standard deviation.
5. Calculate the standard deviation of the mean.
6. Calculate the standard deviation of the standard deviation.
7. (a) Within which limits does a single reading have a 68% chance of falling? (b) Which limits give a 95% chance?
8. Within which limits does the mean have (a) a 68% chance, and (b) a 95% chance of falling?
9. Within which limits does the sample standard deviation stand (a) a 68% chance and (b) a 95% chance of falling?
10. Calculate a value for the constant h in the equation for the Gaussian distribution.
11. If a single reading of 55 had been obtained in the set, would you have decided in favor of accepting it or rejecting it?
12. Take two randomly chosen samples of five observations each from the main set of readings. Calculate their sample means and standard deviations to see how they compare with each other and with the more precise values obtained from the big sample.
13. If the experiment requires that the standard deviation of the mean should not exceed 1% of the mean value, how many readings are required?
14. If the standard deviation of the universe distribution must be known within 5%, how many readings are required?
15. Repeated measurements of the diameter of a wire of circular cross section gave a mean of 0.62 mm with a sample standard deviation of 0.04 mm. What is the standard deviation for the calculated value of the cross-sectional area?
16. The wavelength of the two yellow lines in the sodium spectrum are measured to be 589.11×10^{-9} m and 589.68×10^{-9} m, each with a standard deviation of 0.15×10^{-9} m. What is the standard deviation for the calculated difference in wavelength between the two lines?
17. A simple pendulum is used to measure g using $T = 2\pi\sqrt{l/g}$. Twenty measurements of T give a mean of 1.82 sec and a sample standard deviation of 0.06 sec. Ten measurements of l give a mean of 0.823 m and a sample standard deviation of 0.014 m. What is the standard deviation of the mean for the calculated value of g?

4

Scientific Thinking and Experimenting

4–1 OBSERVATIONS AND MODELS

In this chapter we briefly review the nature of scientific activity in the hope that the different types of experimenting will be seen to arise naturally from the problems that are encountered in various experimental circumstances. In order to understand the nature of scientific thinking, it will help if we go back to fundamentals and pretend that we are inventing a new area of scientific study right from the beginning.

(a) Identification of Significant Variables

As we encounter, through observation, a totally new phenomenon, our natural first question is: what causes this? This question was asked with respect to the diffraction of light, radioactivity, superconductivity, pulsars, and every other physical phenomenon. It is still being asked with respect to the nature of elementary particles, climatic change, cancer, and, of course, many other topics. Asking questions about causes, however, can lead to philosophic difficulties, and it is better to recognize that our natural questions about "explanations" for physical phenomena are really questions about the relationships between observed variables. Normally the first phase of research on a totally new phenomenon consists of a search for variables which seem to be related. By identifying these significant variables, we narrow the field of investigation to practicable levels and facilitate continued work at both experimental and theoretical levels. For example, a valid starting point for a theory of biological evolution might be an assertion about related observations, such as: fossil types are related to the ages of the rocks in which they are found.

It is interesting to notice that, at this primary stage of scientific development, we can make relatively definite statements because we are talking about actual observations. This accounts for the reputation of scientific activities that they lead to "scientific truth" about the universe. The claim must, however, be restricted to the early, diagnostic stage at which we identify the significant variables. Following this stage are later stages in which we deal with a totally different type of activity.

(b) Concept of a Model

After we have analyzed our new phenomenon and are aware of the significant variables, we can proceed to the next level of sophistication. To illustrate this stage let us consider an elementary example. Suppose we were going to paint a wall and wished to know the amount of paint to order. We would have to know the area of the wall, so what would we do? The natural reaction would be to measure the length of the wall and its height and then multiply the two numbers together. But what would that give us? And why would we think that the product has anything to do with the wall? When we multiply these two numbers together, we do obtain something; it is the area of the completely imaginary rectangle defined by these two lengths. This may or may not have any relationship to the wall. The important thing to notice is that we are dealing with two completely different categories. First, there is the real wall whose area we desire. Second, there is a completely invented, conceptual rectangle that is constructed from definitions, exists in our imagination only, and, in the present case, is specified by the two measured lengths. We are commonly insensitive to this very important distinction, because we are all so familiar with the concept of rectangles that a simple, almost subconscious, glance at the wall reassures us that a rectangle is a satisfactory representation of the wall.

But suppose we were not able to make that judgment. Suppose we were blind and had done nothing more than measure the base and one side of the wall without thinking about angles or any other property of the wall. We could multiply our two dimensions to obtain an area which had no relevance at all to a wall whose form happened to be that of a parallelogram. In order to avoid that kind of error, we as blind experimenters would have to remember that the multiplying process gives us the area of a purely hypothetical rectangle, and that this area will be relevant to the wall only to the extent that the hypothetical construct is in correspondence with the actual wall. To check that correspondence, we would have to know the various properties of rectangles, and we would have to compare with the actual wall as many of these properties as possible. We would have to test such properties as straightness of sides, right-angle corners, equality of diagonals, etc. Only after a sufficient number of properties had been compared between the rectangular construct and the real wall, and found to correspond adequately, could we have faith that the area of the imaginary rectangle is a good enough approximation to the actual area of the real wall.

We have now identified a most important distinction. On the one hand we have the real world and our perceptions of it; on the other hand we have hypothetical, imaginary constructs, fabricated out of sets of definitions. Such a construct is often called a **model** of the situation, and the use of models is almost universal in our thinking, whether scientific or nonscientific. The painter has in mind an imaginary rectangle as he contemplates painting the wall. In addition to the real flower in his hand the botanist is aware of the concept of that species, defined by a standardized list of properties. The economist studying the economy of a country constructs a set of definitions and equations whose properties are, he hopes, similar to the actual properties of the real economy. Models give us a framework for thought and communication, a shorthand description for systems, a basis for calculation, a guide for future study, and many other advantages.

Models are of many different kinds, but they have one characteristic in common: they are invented concepts. They are constructed with the intention that they will correspond as closely as possible with the real world, but no model can ever be an exact replica of its real counterpart. They belong to different categories; a wall cannot actually *be* rectangle, nor a wheel a circle. However, the properties of a model may be similar to the properties of the real world, and, in general terms, a model is useful to the extent that its properties do correspond with those of the real world.

(c) Comparison Between Models and the Real World

At the beginning of an experimental study we are usually unaware of the extent to which the properties of our model and its real-world counterpart correspond, and it is necessary, as a basis for all later work, to start by testing the model against the real system. Only if the properties of the model are shown experimentally to be adequately in correspondence with those of the real system are we justified, like the painter about to order paint, in proceeding to the next step.

We must notice in passing that, to be useful scientifically, a model or concept must be actually testable against observation. Thus a proposition regarding the number of angels who can dance on the head of a pin cannot qualify as science. This is not to say that the only useful ideas are those that can be tested against experience, only that other propositions do not come under the heading of science. These other propositions may be perfectly valid as mathematical or philosophical statements, perhaps, or as aesthetic or ethical judgments.

(d) Refinement of Models

In general, an experimental situation will contain, first, the system itself, and, second, a model or models of the system. Whatever else is involved, it will be an essential part of the experimenter's task to test the properties of the model against the properties of the real system. Inevitably our model will be, in principle, incomplete and inaccurate. For example, let us return to the painter and the wall. If he tests the properties of rectangles against the wall with increasing precision, he will almost certainly reach a point at which he begins to find discrepancies. He can thereupon change his model—an angle here, a length there, etc.—to bring it into closer correspondence with the real wall, and the calculated area will be an increasingly accurate estimate of the actual area of the wall. Even with these adjustments, however, the model remains an invented concept, and the area calculated from the model belongs to the model.

In scientific work generally we should feel free to change our models at any time as the need arises. The model is our construct to begin with, and it is only an idea that exists in our heads. In contemplating change, therefore, our only consideration is the basic usefulness of the idea and its improved utility if it is altered in any way. Since it is presumably impossible to construct a verbal or mathematical description of a piece of the natural world that is the exact and total equivalent of the real thing, a process of continued refinement and eventual replacement of models must be accepted as the natural course of events. It is the normal business of scientists, whether "pure," technological, or social, to use the process of comparing models and systems in a continuous search for improvement in models. This is usually not an easy process, because the models we have now are as good as generations of scientists in the past have been able to make them.

On the other hand we need not be totally preoccupied with improving models. Even if no model can be the exact equivalent of the real thing, often the properties of our models and systems correspond sufficiently well for our purposes. If so, we have no need to preoccupy ourselves with the remaining defects. We can proceed confidently with our particular task, provided that we remember periodically to recheck the situation and confirm the continued suitability of the model.

(e) Model Building in the History of Science

It is possible to gain the impression from the above outline that in scientific development there is some kind of unique sequence proceeding from observation to generalization to model. Indeed, scientific thinking quite frequently did proceed in this way. The sequence, however, is not invariable. There are many examples of a basic, invented idea, the foundation of a model, that was the fruit of pure speculation by the originator, without awareness of the observations that could be directly associated with the conjecture. We can recall, for example, de Broglie's speculation on the wave model of matter, which was published in 1924 before any of the relevant phenomena were observed directly, and Fermi's invention of the concept of the neutrino almost forty years before its direct observation. There is no single process of scientific development, no single "scientific method." Ideas and observations tend to shuffle forward roughly together, but with no automatic leadership from one or the other. Regardless of the precise order of development, one point remains invariable: the fundamental activity in scientific experimenting is to compare the properties of models with the corresponding properties of the real world.

We have not discussed at all the processes by which new ideas are introduced as a basis for totally new theories. It is relatively easy to see how an existing idea can undergo a process of continued refinement and attain closer and closer correspondence with observation without any alteration of the basic concepts on which the theory is founded. (Consider, for example, the Ptolemaic theory of planetary epicycles.) On the other hand a theory such as Einstein's theory of general relativity or Schroedinger's wave mechanics can be introduced only after completely radical revision of basic concepts and ways of thinking. Such revision is not easy, partly because our existing ideas tend to shape observing processes in such a way as to promote conformity between concept and observation. A long-established concept, definition, or principle, by its very familiarity, often restricts our vision so that we see only that which we are conditioned to see. This can make it difficult for us to notice the discrepancy, perhaps small but vital, that can stimulate radical revision in thinking. The manner in which such major revolutions in scientific thinking have occurred is described in the books by Kuhn, Cohen, and Harré that are listed in the Bibliography.

One might think that, following such major revolutions, a superseded model or theory would be immediately discarded to make way for its successor. Indeed many models or theories have found no lasting usefulness—one does not hear too much these days about phlogiston, or about earth, fire, air, and water—but this is not always the case. Quite often a superseded model has sufficiently close correspondence with the system that, usually on account of simplicity, it can continue to be very useful, provided we do not expect the increased precision that is available only from a model having a higher order of generality.

(f) Detailed Comparison Between Models and Systems

To summarize our development so far, we have four ingredients in the scientific recipe: (1) observation, (2) an idea constructed in our imagination, (3) the process of comparing the properties of the idea with the real world, and (4) the possibility of modifying the idea progressively to improve the fit between the model and the system. We now turn our attention to the actual procedures by which we can compare the properties of models and systems. To do this it will not be sufficient to have a vague, pictorial concept of the situation; in order to supply an adequate basis for comparison we must be as explicit as possible. This will normally require quantitive observation of the system and mathematical procedures for specifying the model. Let us consider some specific examples and investigate the various levels of sophistication of model construction.

Consider an elastic band, suspended from its upper end, from whose lower end we can hang weights. The most primitive form of construct with respect to the properties of the system would be a verbal description of its behavior. We could say something like: "As I hang more weights at the bottom of the elastic band, it stretches farther." This verbal description and its repeated observation elsewhere could prompt us to invent the general concept "springiness" to serve as a model. But if we wish to refine the model to make it more useful for detailed comparison with observation, our purely verbal methods of description start to fail us; we cannot refine such a vague concept as springiness without resorting to the precision of description that is available in numerical and mathematical modes of expression. We would then make a series of measurements of the extension of the elastic band as a function of load, hoping that they will suggest a more explicit concept. We would obtain a set of measurements such as those shown in Table 4–1. (Note that we pretend to know the weights exactly so that, for simplification, we ignore the uncertainty in them. The lengths are measurements made by us, and so must be expressed in terms of uncertainty.)

Now that we have the measurements, do they give us a complete and adequate description of the results? Not really. It is difficult to judge the behavior of a system from a set of numbers in a table, and some form of visual presentation is much superior. A simple graph of the observations will comprise all the information contained in the table and will, in addition, confer the enormous benefit of facilitating visual judgment of the results. We therefore offer a dia-

TABLE 4–1 Extension vs. Load
for a Rubber Sample

Load, kg	Extension, meters
0.05	0.03 ± 0.01
0.10	0.04
0.15	0.08
0.20	0.13
0.25	0.19
0.30	0.30
0.35	0.34
0.40	0.38
0.45	0.39

gram such as Fig. 4–1, in which we have plotted, in addition to to the central values of the measured variables, the actual intervals over which the measurements of extension were uncertain. Note that we have done nothing more than plot the observations on the graph. At this stage the set of observations is the

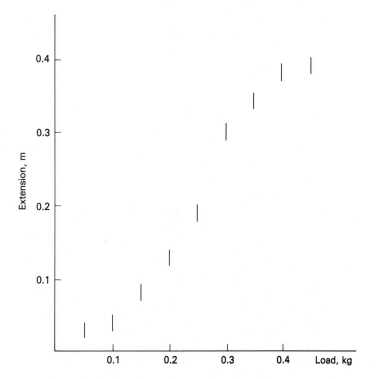

Figure 4–1 Graph of extension vs. load for rubber sample.

only thing we have, and there is no justification for putting anything else on the graph.

This completes the first stage of the process—observation. We must now undertake the next stage in which we construct a model, or models, of the system.

4–2 CONSTRUCTION OF MODELS

The type of process required at the various stages depends very much on the particular experiment. For example, we may be experimenting on a phenomenon for which there are no existing ideas. In such a case our task could be to generate some kind of model. On the other hand we may have some existing proposal or theory that can be applied to our system, thus creating a model. Whatever the circumstances, we shall draw a distinction between models that are "empirical" and models that are "theoretical." By "empirical" we mean a model that is based solely on the observations, without any reference to the detailed, internal operation of the system. The process and its usefulness will be described below. On the other hand a theoretical model is constructed more generally, not just for a particular range of observations, and is based on some concept or principle about the actual mode of operation of the system. The nature of theoretical models and their usefulness will also be described. Let us consider each type in turn.

(a) Empirical Models

In this section we shall assume that we have made a set of observations on a system for which there is no existing model. All we have, therefore, is a set of observations on some property of the system. It could be, for example, the load vs. extension measurements on our elastic band, and the results will probably take the form of a graph like that in Fig. 4–1. Our problem is to construct a suitable model. What can we do? There are several possibilities, and we shall consider them in order of increasing complexity.

(1) Verbal statement.   The simplest model of all would be a simple verbal description of the variation: "The extension increases smoothly with load in an S-shaped curve." Notice that even this simple sentence is a *construct*. As soon as we stop talking about the individual observations and start talking about the variation of the extension with load, we have made the transition from statements about particular observations to constructed presumptions about the behavior of the system. Even such a vague proposition as the statement above could, on closer measurement, turn out to be unsatisfactory. Perhaps, for example, the variation is stepwise instead of smooth. The hypothetical nature of such statements is stressed here as a reminder that we must always be aware of the distinction between statements about the observations themselves and statements that sound as if they were about the observations but are actually statements about our *ideas* concerning the observations.

(2) "Drawing a smooth curve through the points."   The next stage of sophistication in model construction is represented by a process that is so commonly carried out (usually without due regard to its significance) that it is named in the heading for this section. Let us remind ourselves that, as we view the graph of observations initially (Fig. 4–1), it contains the observational points and *nothing else;* we have no basis yet for putting anything else on the diagram. There will inevitably be some scatter in the points because of their inherent uncertainty, but it is possible to base our model construction on the single, basic presumption that, uncertainty and scatter notwithstanding, the actual behavior of the system is smooth and continuous. Note that this is *our* concept or idea, and, as we draw a "smooth curve" (Fig. 4–2) through the points, we are assuming its validity for the system. The presumption of smooth, continu-

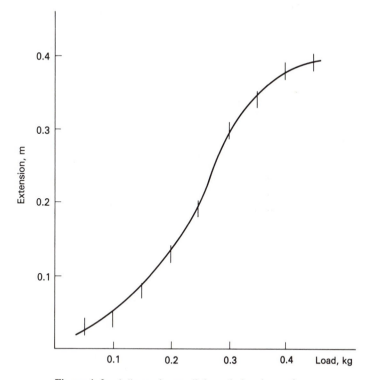

Figure 4–2   A "smooth curve" through the observations.

ous behavior can be valid to a high degree of accuracy for many systems—such as, for example, planetary motion, for which many of the procedures for treating observations were first devised. In all cases, however, the responsibility for deciding to asssume smooth and regular behavior lies with the experimenter, and the assumption should be made only if familiarity with the system leads to the carefully considered conviction that it is valid.

The benefits of assuming regular behavior and drawing a smooth curve through the points can be substantial. One of the most obvious benefits is associated with interpolation and extrapolation. Consider that we have the set of observations shown originally in Fig. 4–1 and that we have drawn a smooth curve through the points as shown in Fig. 4–2. Our knowledge of the system is good at the points at which measurements have actually beeen made. If, however, we desire the value of the extension at a load intermediate between two of our measured values, we have a problem. We could go back to the apparatus and make the desired measurement, but for many reasons this course of action could be either impossible or undesirable. We are then left only with the possibility of *inferring* the desired value on the basis of the existing measure-

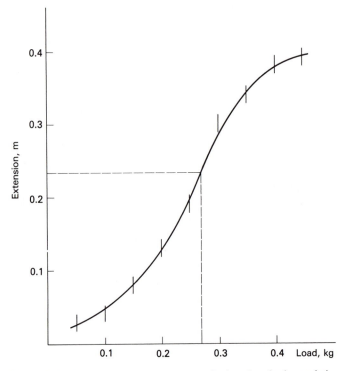

Figure 4–3   The use of a smooth curve to obtain values by interpolation.

ments. The smooth curve we have drawn provides one way of doing so, as shown in Fig. 4–3. We must, however, be careful to remember that the answer obtained by interpolation is an *inferred* value, dependent on our choice of the smooth curve. Likewise, it is possible to use a smooth curve to extrapolate beyond the existing range of values as shown in Fig. 4–4. Such a procedure enables us to make a guess at values outside the measured range, but the validity of the procedure is obviously much more limited than was the case for interpolation. We must have very good reasons for believing that the behavior of the system remains regular beyond the measured range, because smooth variation inside the measured range does not, by itself, offer any guarantees about a wider range of behavior (Fig. 4–5).

Mathematical methods for interpolation and extrapolation will be found in Appendix 3, and they can be used to obtain intermediate values by calculation, without actually drawing the smooth curve. Such methods, however, are still dependent on the assumption of smooth, regular behavior of the system, and the inferred values draw their validity from the reliability of that assumption.

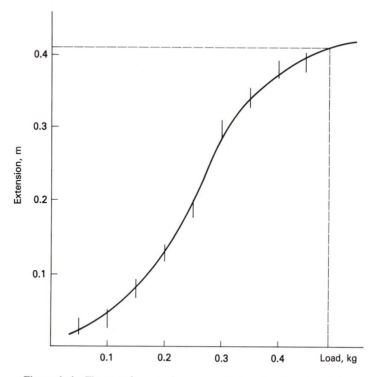

Figure 4–4   The use of a smooth curve to obtain values by extrapolation.

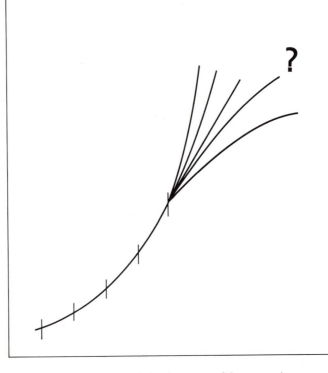

Figure 4–5   Extrapolation is not one of the exact sciences.

Because the validity of interpolation and extrapolation is limited by the assumption of smooth and regular behavior, opportunities for error abound. If, for example, we were to offer someone the graph of temperature vs. time (Fig. 4–6), without specifying the system, and ask them to infer the value of temperature for a time halfway between two measured values, the usual answer is the value shown on the graph. We could then reveal that the graph depicts the noonday temperatures for the first few days of this month and we were asking for a temperature at midnight. Likewise, anyone who has belief in the infallible validity of extrapolation can be asked why he has not made a fortune on the stock market.

Before we leave the topic of drawing smooth curves through points, one final procedure deserves mention. We commonly encounter graphs in which the points have been connected by straight-line segments (Fig. 4–7) or some similar device. How are we to interpret such a diagram? Surely we are not being asked to believe that these segments represent the behavior of the system between the measured points, and the only possible benefit seems to be to sup-

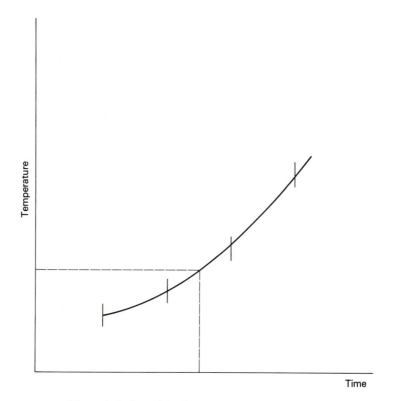

Figure 4–6   Potentially fallacious use of interpolation.

ply some kind of emphasis. Sometimes, in a diagram containing a number of graphs, possibly intersecting, they do help in identifying the various graphs. Since, however, such segments represent satisfactorily neither of the two basic ingredients of experimenting, observations and models, their use is rarely beneficial and they can be misleading. For scientific work they are not recommended except in special cases.

(3) Function finding.   As a more sophisticated form of drawing smooth curves through points we can use a variety of mathematical methods to find an analytical function that will fit the points. Obviously, despite all the mathematical sophistication that appears, such procedures still depend for their validity on the basic presumption of regular behavior in the system; the curves and functions are *our* concept of the behavior of the system. Nevertheless, functions generated empirically to fit sets of observations can be very useful. As mathematical models of the system they can, with varying precision, be used to obtain inferred values for some characteristic of the system. For example,

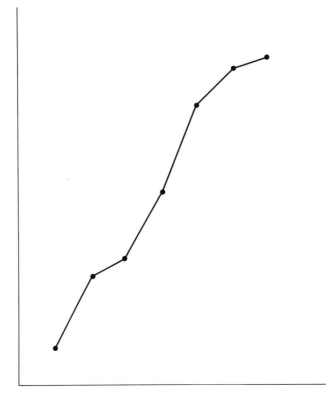

Figure 4–7   A graph showing the use of line segments.

mathematical models of the national economy can be used to evaluate the probable effect on the gross national product of a change in income tax rates, or, more precisely, mathematical models of planetary motion allow us to predict the time of sunrise tomorrow morning or the date of the next eclipse.

It will suffice at present to note the possibility of constructing empirical mathematical models of systems. The methods for doing this will be described in Chap. 6.

(b) Theoretical Models

Theoretical models are part of our familiar, theoretical physics. All analytical theories in physics are constructed out of basic building blocks, definitions, axioms, hypotheses, principles, etc., followed by analytical derivation from these basic starting points. Since all the elements of theories are constructs of human imagination, the theories themselves and the results of the theories are similarly imaginary constructs. Their relevance to actual systems must be evaluated through experiment.

Let us illustrate the situation using a particular example. Consider a system in which we can release a steel ball bearing to fall freely under gravity, and let us measure its time of fall from various heights. If we wished to construct an empirical model of this system, we could simply measure the time of fall over a number of different distances and graph the result, which would look something like Fig. 4–8. We could then use one of the techniques from the preceding section to obtain an empirical model for the system. If, however, we wished to construct a theoretical model of the situation, our approach would be completely different. We would have to choose a set of basic axioms or hypotheses from which we would derive the required results. For example, we might decide to use as a basic hypothesis a presumed value for the acceleration of the ball bearing:

$$a = 9.8 \text{ m sec}^{-2}$$

Notice that this hypothesis already contains several presumptions about the system, thereby starting our process of constructing an invented model. For ex-

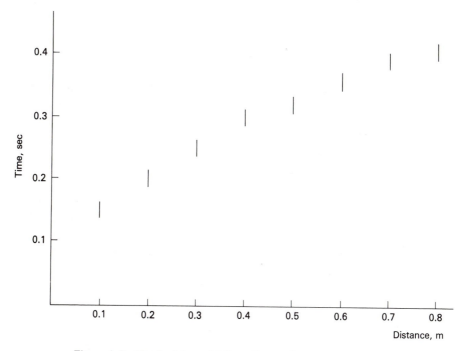

Figure 4–8 Graph of time of fall vs. distance for a freely falling object.

ample, by choosing a constant value for the acceleration we are implicitly neglecting the presence of air resistance. We have every right to do so. Since the
model is ours, we are free to construct it in any way we please. Whether that
presumption makes it a *good* model we may not yet be able to tell. As a second
example, we are also neglecting effects associated with general relativity;
whether this is a serious defect also remains to be seen. We should, of course,
try to estimate in advance the validity of the assumptions that are built into the
model, but we are often limited in our ability to do this. There is always some
point at which we have to decide to start experimenting with the model as it is
and to rely on the experimental results to tell us if further refinement of the
model is necessary.

Assuming that the methods of differential calculus are appropriate to our
problem, we can proceed with the derivation. By integration we obtain:

$$v = 9.8t \qquad \text{assuming } v = 0 \text{ at } t = 0$$

and

$$x = \frac{9.8}{2}t^2 \qquad \text{assuming } x = 0 \text{ at } t = 0$$

or

$$t = \left(\frac{1}{4.9}\right)^{1/2} x^{1/2}$$

In the course of the derivation, all the assumptions we insert constitute further
components of the model. The final result for the measureable variable, the
time of fall, is thus a property of the model. Its applicability to the system must
be the next topic of investigation.

4–3 TESTING THEORETICAL MODELS

Let us consider an actual experiment on free fall under gravity that gave the results shown in Table 4–2. These measurements describe the behavior of the
system. We also have the behavior of the model in the form of the function
that was the outcome of the analytical derivation:

$$t = \left(\frac{1}{4.9}\right)^{1/2} x^{1/2}$$

Our task is somehow to compare these two, but it is not at all clear how that
should be done. One simple suggestion would be to insert the various values of
x in the equation and calculate corresponding values of *t*. We could then compare these with the measured values. If they agreed exactly, we could, per-

TABLE 4–2 Experimentally
Measured Time of Fall vs. Distance
for a Freely Falling Object

Distance, x, m	Time, t, sec
0.1	0.148 ± 0.005
0.2	0.196
0.3	0.244
0.4	0.290
0.5	0.315
0.6	0.352
0.7	0.385
0.8	0.403

haps, be confident that the system and the model were in correspondence. The probability of that happening, however, is minute; apart from anything else, the presence of uncertainty in the measurements would eliminate the possibility of exact correspondence. The major point, though, is that our model is most unlikely to be totally free of systematic defects and deficiencies. It is, in fact, one of our principal purposes in experimenting to detect these discrepancies and deal with them constructively. The possibility of doing this effectively using simple arithmetic comparison is small. Much more significant for our purpose is the overall behavior of the system, and the best way to view this is on a graph.

The graph of our observations, shown in Fig. 4–9(a), consists of a series of points. The graph of the model's behavior is a curve, which is shown in Fig. 4–9(b). The two graphs together give us a visual impression of the relationship between the properties of the system and the model. Our comparison would be more detailed yet, however, if we could pick up one of the graphs and lay it over the other. Doing this, we obtain the composite diagram shown in Fig. 4–9(c). Note that this diagram contains two different components: (1) points representing the properties of the system, and (2) a line corresponding to the analytical function that belongs to the model.

At last we can make a detailed comparison between the overall properties of the system and of the model. By straightforward visual inspection we shall be able to say that the model and the system are in correspondence, divergent, or whatever. We shall list the various possibilities in more detail in Chap. 6. For the present, we must note carefully the kind of statement we are able to make at the end of an experiment. We can say only that the behavior of the model and of the system were in correspondence (or were not) to such-and-such an extent. It is pointless to agonize over a decision that a theory is "true," "correct," "wrong," or whatever. We should avoid using terms like this, even

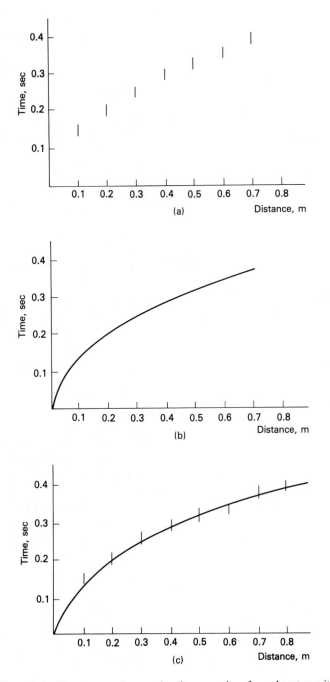

Figure 4–9 The process of comparing the properties of a real system with the properties of a model.

if we are sure that we ourselves are clear about the situation, because of the extensive opportunities for being misunderstood. It is far better to categorize a theory or model as "satisfactory" or "good enough," or some similar phrase, because all such decisions are relative to the purposes we have in mind.

For example, our simple model of constant acceleration under gravity may be perfectly satisfactory for finding the depth of a well by dropping a stone down in it, but it would not be satisfactory for calculating the trajectory of a space vehicle en route to the moon. If that were our purpose, we would have to construct a more refined theory until we had one that is "good enough" for that purpose. Even then we might find that a theory that is adequate for moon rockets is inadequate to describe the motion of the planet Mercury. For that purpose the theory of Newtonian gravitation must be replaced by Einstein's theory of general relativity. As before, the adequacy of Einstein's theory for describing Mercury's orbit (at a particular level of precision) does not "prove" that it is true or correct, simply that it is good enough for that purpose. Equally, the presence of Einstein's theory does not discredit either Newton's theory of gravitation or our simple model of constant acceleration under gravity. Most people do not measure the depth of wells using Einstein's relativity theory. In general we use a particular theory because it is good enough for the purposes in hand. If increased precision is desired at any time, the necessary refinements can be introduced as required (unless, of course, we are working at the limits of knowledge in a particular area, and the chief obstacle is the absence of an improved theory).

Since we are no longer going to proceed using the misleading concept of the "truth" or "falsehood" of theories and models, we shall be dependent on our own decision that a chosen model is good enough, or not, for our purpose. One of the primary aims of experiment design is, therefore, to test the models we use and check their suitability for our purpose. If it is properly planned, our experiment itself will tell us whether our model or theory is good enough.

In passing, we can note one interesting point of philosophy. Even when our system and model seem to be in complete correspondence, we have to be careful about stating the outcome. All we can say is that, at a particular level of precision, we have failed to observe any discrepancy between the system and the model: we cannot claim to have "proved" a theory to be "correct." On the other hand it is possible to be more assertive if we are sure that the properties of the model and system are in disagreement by an amount clearly in excess of the measurement uncertainty; we can say definitely that the model is not in correspondence with the system. If you like, we can say that we have "proved" the theory to be "wrong"—although, even in this case, we might better call it "unsuitable" or "inadequate."

Before proceeding, we must note that this process of comparing systems and models depends on our ability to draw the graphs of the functions that appear in the models. At one time this presented substantial difficulties for even simple functions, like parabolas, and often insuperable difficulties for more complicated functions. Now, however, the use of computers allows us to make the comparison directly. If a suitable computer or terminal is available, we can arrange to have our experimental results displayed directly, either on paper on a plotter or on a video display screen, along with the graph of any function we choose. There are enormous advantages in such use of computers, because it is so easy to change the characteristics or parameters of the model in a search for better correspondence between model and observations. We can also try out completely different functions, thereby greatly extending the range of models that can be compared with the system. Figure 4–10 shows such a plot, in which the computer has offered us two possible graphs, making a choice between the functions extremely easy. If, in addition, we are sufficiently fortunate that we can feed the output of our measuring systems directly into the computer so that continuous processing of observations and comparison with models becomes possible, a whole new domain of experimenting becomes available. The computer will not only appraise our experiment continuously,

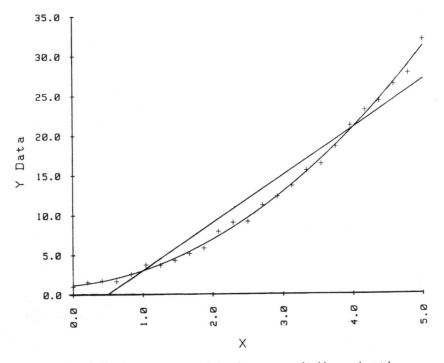

Figure 4–10 Computer-generated functions compared with experimental observations.

permitting ongoing refinement of the model, but also can be used to control the apparatus itself, thereby raising our experimenting to a level not available using hand-powered methods.

Even in this computer age, however, we must not neglect the development of our own personal expertise in experimenting. First, when no computer is available, we are forced to rely on our own resources. Second, even when we are engaged in computer-based experimenting, we stand the chance of obtaining completely meaningless results unless we are totally aware of every detail of the experiment management and computation that has been handed over to the computer. In order to develop our personal expertise, therefore, we must turn our attention to the old-fashioned procedures that constitute the basis of almost all experimenting.

If we are restricted to pencils, graph paper, and rulers, we are virtually compelled to compare the model with the system using the only function whose graph is easy to draw—a straight line. (For fairly obvious reasons we exclude from consideration that other simple graph, the circle.) A large amount of experimental analysis is still carried out in linear form, and, even if the graphical techniques may be cumbersome and tedious in comparison with the ease of computer-assisted plotting, the methods are sufficiently powerful, important, and commonly used that we must become familiar with them.

4–4 USE OF STRAIGHT-LINE ANALYSIS

Our objective is to arrange the plotting process so that the behavior of the system and the model are represented on a graph in linear form.

Consider our former function for the time of free fall under gravity

$$t = \left(\frac{1}{4.905}\right)^{1/2} x^{1/2}$$

which leads to a parabolic representation on a t, x graph. Supposing we were to plot, not t vs. x, but t vs. $x^{1/2}$, our equation

$$t = 0.4515 x^{1/2}$$

could then be compared with the equation for a straight line:

vertical variable = slope × horizontal variable

using

vertical variable = t

horizontal variable = $x^{1/2}$

TABLE 4–3 Experimentally Measured Time of Fall
vs. Square Root of Distance for a Freely Falling Object

Distance, x, m	(Distance)$^{1/2}$, $x^{1/2}$, m$^{1/2}$	Time, t, sec
0.1	0.316	0.148 ± 0.005
0.2	0.447	0.196
0.3	0.548	0.244
0.4	0.632	0.290
0.5	0.707	0.315
0.6	0.775	0.352
0.7	0.837	0.385
0.8	0.894	0.403

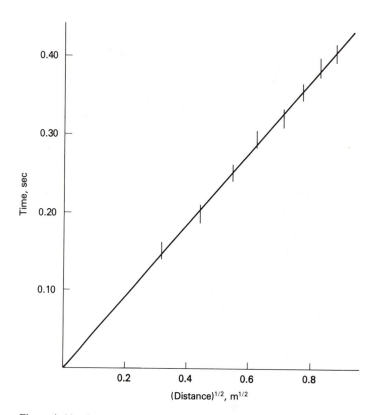

Figure 4–11 Comparison between the properties of a model and of the system when expressed in straight-line form.

and

$$\text{slope} = 0.4515$$

The experimental values of $x^{1/2}$ and t are given in Table 4–3 and are plotted in Fig. 4–11. This graph also contains the line representing the function, and the resulting simplification is immediately obvious. The whole process of comparison is facilitated, and we can identify immediately the degree of correspondence between the model and the system.

Notice that the process would have been equally effective if we had plotted t^2 vs. x instead of t vs. $x^{1/2}$. The slope would have been different, but the opportunity to compare the model and the system would have been equally good.

Finally we note that, in this example, the properties of the model were completely specified and the line on Fig. 4–11 representing the model's behavior is unique. The situation is slightly different if the model contains quantities of which we do not know the value; this will be the topic of the following section.

4–5 CASE OF UNDETERMINED CONSTANTS

Let us consider that we are doing an experiment on a spring to determine the extension under various loads. Suppose that we are aware of a proposal (due to Hooke) that extension x can be considered to be proportional to load W. This proposal,

$$x = \text{const.} \times W$$

constitutes an invented model of the system; let us assume that we wish to test this model against our system. The only trouble is that we probably do not know the value of the constant (the "springiness") to use for our spring. Let us suppose that we have made measurements of extension vs. load and plotted them in Fig. 4–12(a). What are we to do to represent the behavior of the model? The equation

$$x = \text{const.} \times W$$

really represents the infinite set of lines on the $W - x$ plane that have all the values of slope, from zero to infinity, that correspond to the infinite range of possibilities for the value of the constant. Some of these lines are represented in Fig. 4–12(b). What, then, constitutes the outcome of our comparison? Laying one graph on top of the other produces the diagram shown in Fig. 4–12(c) and provides us with the opportunity to choose a line that will be compatible with the experimental points.

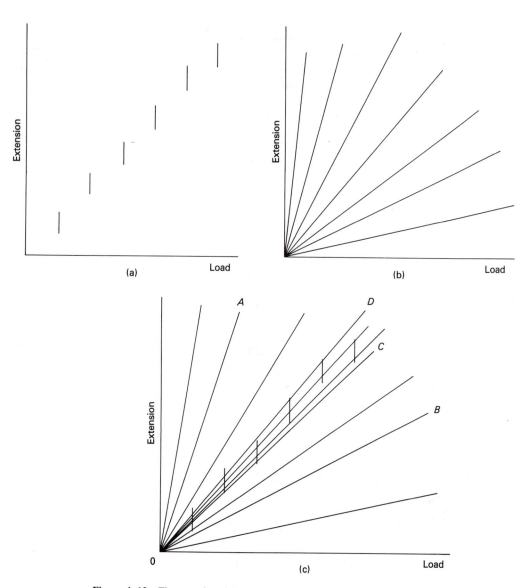

Figure 4–12   The use of straight-line analysis to obtain the value of an undetermined constant in a model.

But which line or lines are we to choose? Clearly the lines *OA* and *OB* have no obvious relevance to the observations and can be disregarded. On the other hand we can identify a certain bundle of lines that fall within the region of uncertainty of the measured points. We can estimate visually the edges of this bundle, and they are represented by the lines *OC* and *OD*. Within these limits all the lines have some degree of consistency with the observations, but no single line stands out as uniquely suitable. All that we can say for the moment, therefore, is that the observations are compatible with the model over a certain range of slopes. This means that there is a certain range of values of "springiness" (which is a property of the model) that are consistent with our system. Our conclusion is, then, that if we have an initially undetermined constant in our model, the experimenting process can be used to determine, within a certain range, the value that is appropriate for our system. This, in fact, is the normal way of determining experimental values of physical quantities.

It is so commonly used because, in addition to the almost necessary use of the graph in comparing models and systems, graphical methods of obtaining values of experimental constants offer so much additional advantage in increased precision that their use is compellingly attractive. The opportunities for error when using algebraic computation alone without graphical checking are great. For example, suppose we are trying to obtain a measured value such as the electrical resistance of a resistor from the variation of the voltage across it with current through it. We would make pairs of measurements of I and V, and we could proceed to use $V = RI$ directly and obtain a value of R from each pair of I, V values by purely algebraic means. We might then hope to obtain an accurate value for R by calculating the average of all the resulting R values.

This approach, however, is deficient in many ways. Basically, of course, it fails to satisfy the primary requirement—to compare the properties of the system and the model—and the consequences for the accuracy of our R value can be serious. If all our pairs of I, V values gave the same, or nearly the same, value for R, we might feel confident in our measurement of R, even without drawing the I, V graph. In the much more likely event, however, that our R values do not all turn out to be the same, we have no way, without the graph, of interpreting the variations.

We might, for example, encounter a case in which, as plotting the graph would have revealed, the points show more scatter than we expected. [See Fig. 4–13(a).] Using a graphical approach, we could still choose a suitable straight line (passing through the origin, if we are sure of the origin as a measured point) and feel reasonably confident about our R value obtained from the slope. Our confidence is justified because the appearance of the graph convinces us that we are dealing with simple scatter about a basically linear variation. Our

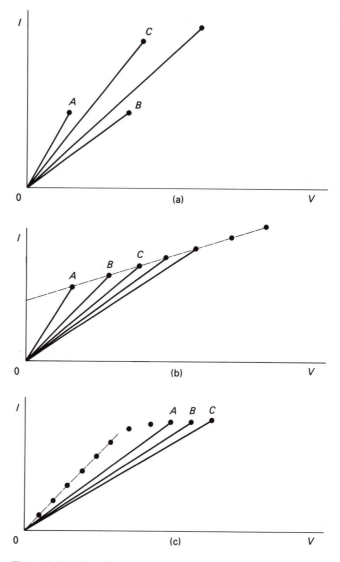

Figure 4–13   Use of graphs to avoid errors in slope measurements.

nongraphical, algebraic calculation would, on the other hand, give us values that correspond to the slopes of the lines OA, OB, OC, etc. On a simple table of values the resulting variability would make no sense at all, and we would gain no insight into what is happening.

As a more significant illustration of the inadequacy of an algebraic, non-graphical approach, consider a case in which some failure of correspondence between the model and the system gives rise either to an unexpected intercept or to deviation from linearity beyond a certain range. Using a graphical approach, we can easily detect and compensate for these discrepancies between model and system. In the first case our graph enables us to obtain a reliable value of R from the slope, which is, in such cases, unaffected by the presence of the intercept. In the second case we would obtain our R value from the slope of the linear portion of the I, V variation, rightly dismissing the nonlinear points as lying outside the scope of the model. Our opportunity to make these judgments can come only from visual inspection of the graph, which makes the situation clear at a glance. Here, as before, nongraphical, algebraic calculation from the pairs of I, V values alone would yield values of R corresponding [see Fig. 4–13(b) and (c)] to the slopes of OA, OB, OC, etc. These slopes have nothing to do with the slope we want, and, if we were to include them in an average of algebraically calculated quantities, we would succeed only in introducing error into our answers.

As we ensure in these ways that our final answer is free from such sources of systematic error, it does not matter whether we know the origin of the discrepancy or not. For the purposes of obtaining the value of the quantity under study it is sufficient at this stage merely to identify the existence of the discrepancy and to ensure that it is not permitted to introduce errors into our answer. We can proceed later to consider possible sources of the discrepancy.

We have discussed the process of obtaining values for initially undetermined constants in terms of slopes only. In principle, since a line has two degrees of freedom on a plane, it is possible to obtain from it two pieces of information independently, such as a slope and an intercept. Because of this, an experiment can be made to yield values for two separate quantities appearing in a theory. Specific methods of doing this in actual practice will be found in Chap. 6.

5

Experiment Design

In Chap. 4 we described the various circumstances in which we compare the properties of models and of systems. We encountered such variety that it will come as no surprise to learn that there is no single way to plan experiments. The techniques and procedures we use depend on circumstances, and we shall describe procedures that are appropriate to a number of cases. The list is not exhaustive, but we shall find that the general principles are valid in a wide variety of experimental circumstances.

We shall assume at the outset that, as the outcome of preliminary investigation, we already know the significant variables. Some of these will be under our control and can serve as input variables; others will take values determined by the system, and they will be the output variables. In the following sections we shall assume that the input variables can be separated and individually controlled; otherwise, with everything varying simultaneously, the interpretation of the results will be much more difficult. This unfortunate circumstance is frequently encountered in professional experimenting, but we shall restrict ourselves to the case of a fully controlled experiment.

5–1 TO TEST AN EXISTING MODEL

In this section we shall be concerned with situations in which a model of some type is already available. This model could be the simplest of suggestions (perhaps even wholly empirical), like $F = kx$ or $V = RI$, or it could be derived from some grand, sophisticated theory like Einstein's theory of general relativity. Whatever the nature of the model, its properties almost invariably will take the form of a functional relation between two or more variables. Our primary objective will be, as always, to compare the properties of the model with those of the system. Only after we have, by experimental test, satisfied ourselves that, over some range at least, the properties of the system and the model overlap, are we entitled to go ahead with the evaluation of the quantity we wish to measure.

Notice that our decision that the model is satisfactory, or not, must be based on the experiment itself. We are, of course, not going to attempt to decide on such meaningless questions as; "Is the model or theory "true" or "false"? "Correct" or "incorrect"? or whatever. As we have said so often, all models are imperfect in principle, and we simply need to know if the model is "good enough" for our purposes at our level of precision. Only our own experiment can provide the basis for making that decision, and we must try to ensure, by careful design, that it does so. Once we have checked that our model is "good enough," we can proceed to evaluate our unknown quantity, while not forgetting that, if our situation changes and increased precision is called for, we must reopen the question of the model's adequacy for our purpose.

As we have described in Sec. 4–3, we shall find, almost invariably, that the best ways of testing models of physical systems involve a graphical approach. In principle we wish to draw a graph of the model's behavior and superimpose on it our observations of the behavior of the system. In order to do this in simple form, however, there are some requirements that we must meet.

First, since a graph (as we are considering it) is a two-dimensional diagram, we must limit ourselves initially to two variables. In many cases this will automatically be satisfied, as it has been in all our preceding examples. In others, however, our output variable will be a function of two (or more) independent input variables. We cannot plot three variables as coordinates on a two-dimensional piece of graph paper (although three-dimensional diagrams can easily be generated by computers and are frequently seen in the scientific literature). Consequently, for our purposes it is necessary to simplify the experiment by holding one of the input variables constant while studying the dependence of the output variable on the other. We can then alter the first variable to a second fixed value and repeat the process. By a succession of such measurements we can build up a relatively complete picture of the behavior of the system. Notice that the success of the process depends on our primary assumption that it is possible to hold one of the input variables constant, independently of variation of the other. If such isolation of the input variables is not possible, we have problems; some of the necessary techniques will be mentioned in Sec. 5–2(b).

For our present purpose let us suppose that we have only one input variable, either because only one exists or because we can isolate one by holding the others constant. Our experimental procedure is clear; we must measure the variation of the output variable with the input variable and plot it for comparison with the corresponding graph for the model. As we have suggested, however, in Sec. 4–3, it would take a computer to draw even simple nonlinear functions, and the advantages of drawing our graphs in straight-line form are so overwhelming that we shall consider only this approach.

5–2 STRAIGHT-LINE FORM FOR EQUATIONS

(a) Simple Cases

If the model we are considering contains only linear functions (such as distance travelled at constant velocity as a function of time, or the potential difference across a constant resistor as a function of current), we have no problem; the equation is already in straight-line form. This is rarely the case, however, and we are almost invariably faced with the necessity of converting the functions found in the model into linear form. We have already encountered this requirement in Sec. 4–3. There the function was

$$t = 0.4515x^{1/2} \qquad \text{(in units of meters and seconds)}$$

and, clearly, if we wish to represent this equation in the linear form,

$$\text{vertical variable} = \text{slope} \times \text{horizontal variable} + \text{constant}$$

we must choose

$$\text{vertical variable} = t$$
$$\text{horizontal variable} = x^{1/2}$$
$$\text{slope} = 0.4515$$

and

$$\text{constant} = 0$$

This is a simple case, and it is often less easy to see how an equation can be converted into linear form. There are no definite rules for doing it. The best way is to keep clearly in mind the form toward which we wish to work,

$$\text{vertical variable} = \text{slope} \times \text{horizontal variable} + \text{constant}$$

and juggle the quantities in our original equation around until we have the required form. Opportunities for practice will be found in the problems at the end of this chapter.

Notice that there is no unique answer in this process. A given equation can sometimes be put in linear form in several different ways. For example, the equation

$$t = 0.4515x^{1/2}$$

can be used equally effectively in any one of the equivalent forms

$$x^{1/2} = \frac{1}{0.4515}t, \qquad t^2 = \frac{1}{4.905}x, \qquad x = 4.905t^2$$

with appropriate choices for *vertical variable, horizontal variable,* and *slope.* There is a certain conventional tendency to plot graphs with the input variable horizontally and the output variable vertically, but there is no real requirement to do so. We should choose the form of graph that most effectively serves our purpose.

Our purpose should include not only the basic experimental requirements (testing models, etc.) but also the comfort and convenience of the experimenter. For this, one should plot variables as simply as possible. For example, consider an experiment to determine a coefficient of viscosity by studying the flow of liquid through a pipe. The appropriate equation (Poiseuille's equation) is

$$Q = \frac{P\pi a^4}{8\eta\ell}$$

where Q = rate of flow
$\quad\;\; P$ = pressure difference across the pipe
$\quad\;\; a$ = pipe radius
$\quad\;\; \ell$ = pipe length
$\quad\;\; \eta$ = coefficient of viscosity

In this case one possible choice would be to plot Q vs. $(\pi a^4/8\ell)P$, a graph whose slope would be equal to $1/\eta$. This would, however, be an unwise choice for a number of reasons. First, it greatly increases the amount of arithmetic required to do the plotting, because each value of P must be multiplied by $\pi a^4/8\ell$. Second, each of the quantities a and ℓ has an attached uncertainty; if this were to be combined each time with the uncertainty in P, we would have a falsely enhanced uncertainty for the compound quantity (for example, a would be measured only once, and its uncertainty should not be combined with that of P as if, every time we measured P, another measurement of a were made to combine with it). Clearly in this case the path of wisdom is to plot Q vs. P and to use $\pi a^4/8\eta\ell$ as the slope, thus avoiding all the difficulties mentioned above. In general it is best to plot variables that are as simple as possible and to leave as much of the arithmetic as possible to be done just once in calculating the answer from the slope.

(b) Use of Compound Variables

In many cases it may suit our convenience (or else it may be absolutely necessary) to use compound variables. Consider, for example, the so-called
"compound pendulum," a rigid lamina of a certain shape which is allowed to
oscillate under gravity about an axis perpendicular to the plane of the lamina
[Fig. 5–1(a)]. The normal model of its oscillation (for small angles) gives the
period T as

$$T = 2\pi\sqrt{\frac{h^2 + k^2}{gh}}$$

where T = period of oscillation (output variable)
  h = distance from center of mass to point of support (input variable)
  g = gravitational acceleration (constant and unknown)
  k = radius of gyration about center of mass (constant and unknown)

Straight-line forms of this equation are not immediately obvious, but it is
clearly impossible to place it in the required linear form if we choose x and y to
be functions of h and T singly. Conversion into linear form using compound
variables is, however, possible. Squaring both sides of the equation, we obtain

$$T^2 = \frac{4\pi^2(h^2 + k^2)}{gh}$$

Therefore

$$T^2 h = \frac{4\pi^2(h^2 + k^2)}{g}$$

and

$$h^2 = \frac{g}{4\pi^2}T^2 h - k^2$$

which is linear with

$$\text{vertical variable} = h^2$$

$$\text{horizontal variable} = T^2 h$$

$$\text{slope} = \frac{g}{4\pi^2}$$

and

$$\text{intercept} = -k^2$$

This example is worth studying, since it illustrates very clearly the supe-
riority of linear analysis over other methods. A commonly encountered ap-

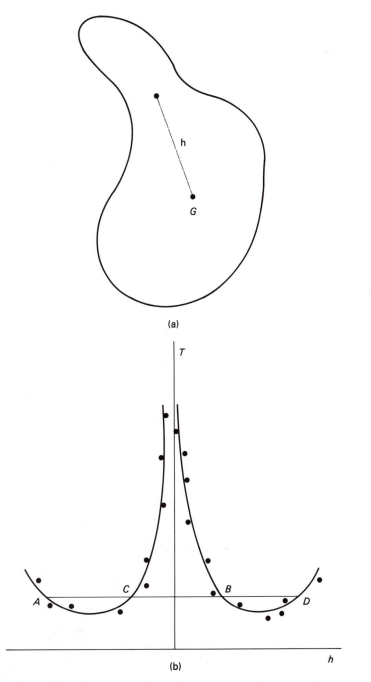

Figure 5–1   The "compound pendulum" and its variation of oscillation period with h.

proach to this experiment uses the graph of T vs. h, which is shown in Fig. 5–1(b). It turns out that k can be obtained from the lengths of the intercepts AB and CD; if g is required, it has to be obtained as a calculation from this value of k. The advantages of the linear form of analysis are clear. First, the T vs. h graph gives no basis for comparing the system with the model, unless one uses a computer to draw the graph of the function $T(h)$. Second, no reliable estimate of the uncertainty of the final answer can be obtained from this graph, while, on the other hand, the overall uncertainty can readily be obtained from the linear graph. Third, the use of an intercept at such a low angle, as illustrated in Fig. 5–1(b), is very unreliable, since small changes in placing the lines can cause large changes in the length of the intercepted portion; a linear form enables us to determine the slope of the graph very reliably. Fourth, the answer using the intercept method is determined solely by a few points in the vicinity of the intercepts, and we obtain no benefit from all the other points; when drawing a straight line, however, all the points can contribute to the choice of the line. Last, the linear graph gives g and k from almost independent measurements on the graph, while, with the other method, any inaccuracy in the value of k is propagated automatically into the value of g.

The use of compound variables can also be convenient when there are two or more separate input variables. In such cases, even if the use of compound variables is not absolutely necessary for linear plotting (as was the case for the compound pendulum), they often provide the neatest and most effective way of plotting our graphs. It was mentioned earlier that, if a system involved two independent input variables, we could study the variation of the output variable with either input variable in isolation, while holding the other input variable at a number of discrete levels. For example, if we wish to measure the specific heat of a fluid by continuous-flow calorimetry, we can allow it to flow at a certain mass flow rate, m, through an electrically heated tube in which the rate of heat generation is Q per unit time. The equation for the resulting heat balance (neglecting losses, etc.) is

$$Q = mC\,\Delta T$$

where C is the specific heat of the fluid, which we wish to measure, and ΔT is the difference in temperature between the input and output ends of the tube. Clearly both Q and m are separately controllable, and we can perform our experiment by studying the variation of ΔT with m, holding Q at various fixed levels, or we can study the variation of ΔT with Q, holding m at various fixed levels. We would then be able to plot either ΔT vs. $1/m$, in which case the various slopes would have values Q/C, or ΔT vs. Q, giving the various slopes as Cm. Another possibility exists, however. If we treat the product $m\,\Delta T$ as one variable and plot it against Q, we shall obtain one graph that summarizes all our information about the system by incorporating both input variables simultaneously, whether we did or did not control the values of m. In this case our slope would simply have the value C, and we would have a neat way of testing the model and obtaining our unknown in one simple step.

Such use of compound variables is very common, and, as before, the choice of combination for the variables and the mode of plotting can be made to suit both the convenience of the experimenter and the requirements of the experiment. If any difficulty in interpreting observations appears when they are plotted using compound variables (unexpected scatter, perhaps, or some systematic deviation from linearity), we can always obtain extra information about the system by plotting pairs of variables individually rather than in combination. This frequently helps in identifying the cause of any difficulty.

(c) Logarithmic Plotting

It is frequently desirable, and sometimes absolutely necessary, to plot variables in logarithmic form. For example, many physical processes involve exponential functions of the form

$$y = ae^{bx}$$

where y and x are measured variables and a and b are constants whose values are to be obtained from the experiment. The equation can be put in linear form by taking logs of both sides to the base e.

$$\log_e y = \log_e a + bx$$

Thus, if we plot $\log_e y$ vertically and x horizontally (known as a "semi-log plot"), our model will give us a straight line. The slope will give us a value for b, and the intercept will give the value of $\log_e a$. Note that, if logs are taken to the base 10 instead of e, only the intercept is affected, and this can be convenient if we are interested only in the slope.

The use of such logarithmic plotting is very common because of the frequent occurrence of exponential functions in the models of physical and chemical processes. In addition, however, logarithmic plotting is used even for simple algebraic functions. Consider, for example, the function

$$y = x^2$$

Taking logs of both sides, either to base 10 or base e, we obtain

$$\log y = 2 \log x$$

This equation is linear with

$$\text{vertical variable} = \log y$$

$$\text{horizontal variable} = \log x$$

$$\text{slope} = 2$$

Thus, functional dependence like a simple square can be tested using such logarithmic plotting (known as a "log-log plot").

But what would be the advantage of this type of graph over a plot, as we have been recommending all along, of y vs. x^2? One obvious answer is that it allows us to plot, on one piece of paper of reasonable dimensions, variations that are too extensive for traditional plotting. A range of one power of 10 in our observations can be conveniently plotted on simple graph paper, a range extending over a factor of 100 is becoming difficult, and a factor of 1000 makes satisfactory plotting impossible. For these very large excursions of our variables, only logarithmic plotting allows realistic representation of the results.

A second advantage of logarithmic plotting concerns the power of the function. If our system is behaving in such a way that the function

$$y = x^{1.8}$$

would be a better model than $y = x^2$, that fact would probably elude us if we plotted y vs. x^2. We would simply obtain a set of points that deviate from a straight line, and the source of the discrepancy would not be immediately obvious. The log-log plot, on the other hand, would still give us a straight line. This would tell us that a function involving a power was still a good model. The slope of the line would not be 2, of course, and the improved value of the exponent, 1.8, would be available from the slope of the log-log graph. We shall consider further (in Chap. 6) the uses of log-log plotting for the construction of empirical models involving such powers. It will suffice for the moment to note that, at the stage of designing an experiment, the possibility of semi-log or log-log plotting should be kept in mind if either the type of function or the range of variables suggests that it is appropriate.

5–3 EXPERIMENT PLANNING

We shall now list the actual, practical steps by which we prepare to do the experiment. These may seem tedious to one whose ambition is to get on with the experiment as quickly as possible and worry later about what to do with the results. Indeed, for many of the simple experiments commonly encountered in teaching laboratories the painstaking care that we are about to recommend may seem pointless and pedantic. But we must remember that the simple experiments in teaching laboratories are merely simulation, in suitably simplified form, of the much more complicated and important situations that we shall encounter later in real systems. If in an introductory physics laboratory we forget to measure the wire diameter while doing an elasticity experiment, it is probably not going to matter too much; we can return to the laboratory later and recover the offending measurement and, even if we do not, the world will not come to an end. If, however, ten years later we are planning some X-ray astronomy in a spacecraft and we notice only when our experiment is in orbit that we have forgotten to measure some vital characteristic of our detector, we shall have wasted someone's very expensive rocketry and we shall be less than popular. We should acquire as early as possible the habit of meticulous and painstaking planning of our experiments, even if, for the moment, it may sometimes seem superfluous.

The planning steps are as follows:

(a) Identify the System and the Model

This may seem somewhat trivial, but it is better to be clear at the beginning about the topic of our experiment. In our falling-ball experiment, for example, are we going to worry about air resistance or not? If we decide to ignore it, we are not being irresponsible—we are merely defining one aspect of our model. Whether this is a good decision will be made clear later by the experiment itself. If the behavior of the system turns out to be in correspondence with the behavior of the model at the level of precision we use, we can be satisfied that there would have been no point in wasting time over small effects. If we make a poor choice, our experimental results will very quickly inform us of the necessity of reconsidering the matter. So at the outset we decide on the limits of the system and the model, and we proceed to test the situation.

(b) Choose the Variables

Usually one quantity in the experiment will be an obvious choice for an output variable. If there is only one input variable, there is no problem. If there are several input variables, we should try to identify one as the chief independent variable and vary the others in discrete steps.

(c) Rectification of Equation

The equation representing the behavior of the model must now be put into straight-line form as described in Sec. 5–2. As we have already mentioned, there is no unique, correct choice for the straight-line form. We choose a form that suits our purposes conveniently and effectively. For example, if our equation contains some unknown quantity whose value is to be determined by the experiment, it is probably best to use a form for the straight line that puts the unknown into the slope. It is, of course, possible to determine unknown quantities from intercepts, but, because intercepts can frequently be subject to errors arising from instrumental defects or other systematic errors, it is usually preferable to obtain our unknowns from slopes. If the equation contains two unknowns, it is probably best to find a form that enables us to obtain one unknown from the slope and the other from an intercept.

(d) Choice of Range of Variables

Before starting the actual measurements, we should make decisions about the ranges over which we hope to make them. It is usually best to plan on a range for the input variable of at least a factor of 10. More is better, and less can often give an unsatisfactory basis for comparing the behavior of systems and models. Obviously we cannot choose directly the range of the output variables; the system itself will tell us these values. We must, however, still be careful. There may be instrumental limits beyond which damage can occur—e.g., elastic limits, overheating of precision resistors, overloading of meters and other instruments. Carefully made trial measurements will allow us to determine the range of the input variables that will allow us to avoid overloading any part of the system. This is the time to consider carefully all aspects of instrumental ratings. For example, does the resistance box have marked on it the maximum allowable current for each range? If so, we incorporate it into the choice of range for the variables. If not, we go and find the values in the catalogue. In all cases we should ensure that limits are identified and observed. It will be too late when the smell of overheated insulation and a vertical column of blue smoke alert us to the frailties of physical apparatus and the expense of restitution.

(e) Choice of Precision of Experiment

We should not start an experiment until we have a general idea of the precision we hope to attain in the overall result. This is not to say that we can guarantee a final level of precision, but we should have a target figure to serve as a guide for our choice of measurement methods. For example, the request "Measure the acceleration of gravity using a pendulum" is, by itself, virtually meaningless. In response to this request we could spend ten minutes with crude apparatus and obtain a result with a precision of 10%, or we could spend weeks with refined, expensive equipment and attain 0.01%. We can obtain a realistic impression of the expectation only from some request like "Measure g using a simple pendulum to within 2% and try not to spend more than two hours on it." The figure 2% gives us a general idea of the kind of measurement we are being asked to make.

We should have such a target in mind for every experiment; it will serve as a basis on which we can proceed to design our experiment realistically. We can try to ensure that, on the one hand, all our measurements are of sufficient precision to contribute usefully to the final result and, on the other hand, we do not waste time and effort making some measurement with precision far in excess of that required.

To see how such a design could be carried out, let us return to the example of the pendulum and our hope for 2% uncertainty in the g value. We know that the result for g, although it will be obtained graphically, in essence involves measurements of ℓ and T (in the form T^2). If the uncertainty in any measurement of either ℓ or T^2 is in excess of 2%, there is little chance that it will contribute usefully to a determination of g within 2%. Suppose, as a first guess, we elect to restrict the uncertainties in each of ℓ and T^2 to fall below 1%; what are the implications for the measurements of ℓ and T? Our first step must be, by making trial measurements, to assess the absolute uncertainty with which we can make our measurements of ℓ and T. Once we have determined these uncertainties, we can find the limits on the ranges of the ℓ and T measurements that allow the precision to be acceptable. Let us assume that, with the apparatus available, we think we can measure lengths with an absolute uncertainty of ± 1 mm. What is the length at which this gives precision of 1%?

$$\frac{0.1}{\ell} = 0.01 \qquad (\ell \text{ in cm})$$

Therefore

$$\ell = 10 \text{ cm}$$

Thus, so long as our lengths are greater than 10 cm, the contribution to the overall uncertainty from ℓ is within acceptable limits, and we have identified one limit on the acceptable range of ℓ.

What are the implications for T? If we are going to ascribe a precision of 1% to T^2, we need 0.5% in T. The period of oscillation will be determined by timing a specified number of oscillations with some kind of timer or stopwatch. Let us suppose we are using a stopwatch that (as we can determine by actually trying the measurement) allows us to measure the time for a number of oscillations to within ± 0.2 sec. Note that this figure of 0.2 sec must be the overall uncertainty in the whole timing process, including the personal judgment of the pendulum's position for pressing the button, response times in reacting, etc.; it is not sufficient to consider solely the reading uncertainty of the stationary hand of the stopwatch. In any case, if we have an overall uncertainty in the timing measurement of ± 0.2 sec, we can calculate the precision of any timed interval, t, as $0.2/t$. This is the quantity that we wish to restrict to values below 0.5%, and so the limiting condition is

$$\frac{0.2}{t} = 0.005$$

The limiting value of t is therefore given by

$$t = \frac{0.2}{0.005} = 40 \text{ sec}$$

Thus, provided we choose the number of pendulum oscillations so that we are never measuring times less than 40 sec, we can hope that our measurements will all contribute effectively to an overall determination of g within 2%. We cannot, of course, guarantee that such an experimental plan will result in a value for g with an uncertainty no greater than 2%; there is always the possibility of unexpected contributions to measurement uncertainty or of unsuspected systematic error. But at least we can avoid making measurements that stand no chance at all of contributing usefully to the overall result.

At this stage we must consider whether each measurement is going to be considered in terms of an estimated range of uncertainty, or whether random fluctuation is large enough to require the use of statistical methods. If the latter, some trial measurements will allow us to make a preliminary estimate of the variance and so enable us to choose the sample size that will be needed in order to attain the required precision. At this stage we should recall the warnings about inaccuracy in small samples that were given in Sec. 3–11. In addition, however, we must remember that attempts to improve precision by increasing the sample size can be unrewarding. The expression for the standard deviation of the mean involves $\sqrt{N}$, so that if a trial sample of 10 measurements suggests the desirability of, say, a tenfold increase in precision, the sample size would have to be increased by a factor of 100. A sample size of 1000 may not be feasible, and we would have to seek some other route to improved precision.

Whether the measurements are statistical in nature or whether they have simple, estimated uncertainty, it will be possible to decide at this stage if each measurement in the experiment can be satisfactorily made or not. If it appears that some measurement is restricted to an uncertainty in excess of our design aspirations, we must either obtain a more precise method of measuring that particular quantity, or, if that is not possible, we must acknowledge that our former target for the overall uncertainty of the experiment was unrealizable with the apparatus available and that revision of the target value will be necessary. Also, by assessing the contribution of each quantity in the experiment to the overall uncertainty, we shall be able to identify any measurement (or measurements) that makes a dominating contribution to the final result, either because of low intrinsic precision or else because of the way in which it enters into the calculation (e.g., some quantity raised to a high power, or a quantity that has to be obtained as the difference between two measured values). Once identified, these measurements can be given special attention so that their uncertainty may be kept under control as much as possible.

All the detail described in this section may seem to constitute an unnecessarily exacting approach to a small, simple experiment, but it is wise to recall once again that we are practicing for much bigger, more important experiments in which the consequences of failure to plan properly can be serious and expensive.

(f) Construction of Measurement Program

After choosing the variables, ranges, and precision of measurement, it is best to conclude the design of the experiment by constructing a complete and explicit measurement program. This will normally take the form of a table that includes all the quantities to be measured in the experiment and also provides space for any computations required for drawing the graphs. A completed measurement program will allow the experimenter, during the course of the experiment, to concentrate on the actual conduct of the experiment. While manipulating apparatus and making measurements, there is usually enough to be done without the continuous necessity of deciding what to do next. The presence of the measurement program will also help to guard against the accidental omission of some significant measurement, which could be overlooked as a consequence of the pressure of actual experimenting.

As we have remarked frequently, all this planning may seem like an unnecessary amount of fuss for a simple experiment. These recommendations, however, represent nothing more than the basic minimum of preparation for any serious experimenting, and no opportunity should be lost for the early formation of careful habits of experiment design and planning. It is important to avoid the temptation to rush ahead with the experiment, leaving until later the task of deciding what to do with the results; it is much more beneficial to acquire the habit of setting aside the time to design and plan an experiment properly before starting the actual measurements.

5–4 EXPERIMENT DESIGN WHEN THERE IS NO EXISTING MODEL

This situation appears when, for example, we are making observations on some phenomenon that is so new that a theoretical model has not yet been constructed, or else on some system that is so complicated, such as a complex engineering system or some aspect of national economics, that it will probably never be possible to construct a theoretical model for it. If we do not have an existing model to test, our objective in doing experiments on the system could take several different forms. We may be motivated by simple curiosity or by a practical need for information about the system. At a higher level we may be interested in the possibilities of model building. We may be seeking guidance for the construction of a theoretical model, or, if this is too difficult, we may wish to obtain measurements to serve as a basis for a purely empirical model of the system. As was mentioned in Chap. 4, empirical models, even in the absence of detailed, theoretical understanding, are extremely useful. They can be helpful in systematizing our thoughts about a complex system, and they are usually essential for such mathematical calculations on the system as interpolation, extrapolation, forecasting, etc.

Whatever our motivation, we would probably like to find a function or graph that provides a good enough fit to the observations. The methods of finding suitable functions will be described in Chap. 6, and we shall restrict ourselves for the moment to the question of designing the experiment. In the absence of an existing model, experiment design can be relatively straightforward, especially if we can isolate the input variables so that we can vary one while holding the others at fixed values. Our experiment design will consist simply of measuring the output variable over suitable ranges of the input variables in order to build up as complete a picture of the behavior of the system as possible. If we cannot isolate the input variables, of course, we have problems, and this case will be considered in Sec. 5–7.

Even if we have no existing theory for a phenomenon, it is wise to accept any available hints about functions that might be appropriate to our system, and to test these possibilities against the system's behavior. One way of obtaining such suggestions will be discussed in the next section.

5–5 DIMENSIONAL ANALYSIS

Even if no complete theory of a physical phenomenon exists, it is still possible to obtain very useful guidance for the performance of an experiment by the method of dimensional analysis. The "dimensions" of a physical (mechanical) quantity are its expression in terms of the elementary quantities of mass, length, and time, denoted by M, L, and T. Thus, velocity has dimensions LT^{-1}, acceleration LT^{-2}, density ML^{-3}, force (equals mass $\times$ acceleration) MLT^{-2}, work (equals force $\times$ distance) ML^2T^{-2}, etc.

The principle used in dimensional analysis is based on the requirement that the overall dimensions on the two sides of an equation must match. Thus, if g is known to be related to the length and period of a pendulum, it is obvious that the only way in which the LT^{-2} of the acceleration on the left side can be balanced on the other side is to incorporate the length to the first power (to give the L) and the period squared (to provide T^{-2}). We can thus say immediately that, whatever the final, theoretical form for the equation, it must have the structure

$$g = \text{(dimensionless constant)} \times \left(\frac{\text{length}}{\text{period}^2} \right)$$

Note that the process can give no information about dimensionless quantities (pure numbers such as π, etc.), and so we must always include their possible presence in equations obtained by dimensional analysis.

The general method is as follows. Consider a quantity z that is assumed to be a function of variables x, y, etc. Write the relation in the form

$$z \propto x^a y^b \ldots$$

where a and b represent numerical powers to which x and y may have to be raised. We then write down the dimensions of the right-hand side in terms of the dimensions of x and y and of the powers a and b. Second, we set down the condition that the total power of the dimension M on the right-hand side must be the same as that known for z. We do this also for L and T, obtaining thereby three simultaneous equations that enable us to calculate values for a, b, etc.

For example, consider the velocity v of transverse waves on a string. We might guess that this velocity is determined by the tension T in the string and the mass per unit length m. Let us write

$$v \propto T^a m^b$$

The appropriate dimensions are:

of v:	LT^{-1}
of T (force):	MLT^{-2}
of m (mass per unit length):	ML^{-1}

Therefore,

$$LT^{-1} = (MLT^{-2})^a (ML^{-1})^b$$
$$= M^{a+b} \times L^{a-b} \times T^{-2a}$$

Thus, by comparing in turn the powers of M, L, and T on the two sides of the equation, we obtain

$$\text{for } M: \qquad 0 = a + b$$
$$\text{for } L: \qquad 1 = a - b$$
$$\text{for } T: \qquad -1 = -2a$$

of which the solutions are obviously

$$a = \tfrac{1}{2}, \qquad b = -\tfrac{1}{2}$$

We obtain finally

$$v = (\text{dimensionless constant}) \times \sqrt{\frac{T}{m}}$$

Such a procedure is very valuable, for, even in the absence of a detailed, fundamental theory, it provides a prediction regarding the behavior of the system. This can be a starting point for experimental investigation, and if the experiment shows consistency between the system's behavior and the model produced by dimensional analysis, we shall have confirmation of the validity of our original guess regarding variables. If the experiment shows a discrepancy, we must look again at our primary suppositions about the quantities involved in the experiment. Notice that in the above example we obtained three equations for only two unknowns. The situation, therefore, was really overdetermined, and we were fortunate that the equations containing a and b were consistent. Had they not been, we would have known immediately that our guess regarding the constituents of v was wrong.

Powerful as this method is, difficulties will obviously arise when the quantity under discussion is a function of more than three variables. We then have more than three unknown powers but only three equations from which to determine them. In this case a unique solution is not possible, but a partial solution may be found in terms of combinations of variables.

For example, consider the flow rate Q of fluid of viscosity coefficient η through a tube of radius r and length ℓ under a pressure difference P. All of these quantities are clearly significant in determining the flow rate, and so we may suggest a relation

$$Q \propto P^a \ell^b \eta^c r^d$$

The dimensions of the quantities are as follows:

Q (volume per unit time): $L^3 T^{-1}$

P (force per unit area): $MLT^{-2} \times L^{-2} = ML^{-1}T^{-2}$

ℓ (tube length): L

η (viscosity coefficient, defined as force
per unit area per unit velocity gradient):

$$(MLT^{-2})(L^2)^{-1}(LT^{-1} \times L^{-1})^{-1}$$
$$= ML^{-1}T^{-1}$$

r (tube radius) L

Therefore

$$L^3T^{-1} = (ML^{-1}T^{-2})^a L^b (ML^{-1}T^{-1})^c L^d$$

Comparing powers of

M: $0 = a + c$

L: $3 = -a + b - c + d$

T: $-1 = -2a - c$

Here we have four unknowns and only three equations, so that, in general, a complete solution is not possible. We can obtain part of it, however, for it is obvious that the M and T equations give us

$$a = 1$$
$$c = -1$$

The equation for Q must therefore contain the term P/η. The remaining part of the solution can be written only as

$$b + d = 3$$

If we write this

$$d = 3 - b$$

we can see that Q must contain the product r^3/r^b. It also contains ℓ^b, so that we can write

$$Q \propto \frac{P}{\eta} \times r^3 \times \left(\frac{\ell}{r}\right)^b$$

Since it is inconceivable that Q should increase with ℓ, if all other quantities are kept constant, it is obvious that b must have a negative value, and we can invert the ℓ/r term to obtain, finally,

$$Q \propto \frac{P}{\eta} \times r^3 \times \left(\frac{r}{\ell}\right)^b$$

The quantity b remains unknown, and this is as far as dimensional analysis can take us toward the complete solution. However, even this partial solution could serve as a guide to experimenting in a situation in which no fundamental theory existed. Dimensional analysis can be extended to cover thermal and electrical quantities, but in those cases ambiguities arise and they require special consideration. The appropriate discussion will be found in the standard texts on heat and electricity or in the specialized texts on dimensional analysis.

5–6 DIFFERENCE-TYPE MEASUREMENTS

In all the preceding sections we have assumed that there was a clear and definite relationship between the input and output variables, and that the input variables themselves were readily identifiable and relatively well controlled. We do, however, encounter circumstances in which we are not so fortunate. Perhaps our input variables cannot be clearly isolated, so that, with everything varying at once, it is difficult to identify the effect of each on the output of the system. Or perhaps the system is so complex and subject to so many variable factors that we find it hard to judge whether the effect in which we are interested even exists. Many experimental techniques, mostly of a statistical nature, have been devised for use in such circumstances. Descriptions of these will be found in the texts on statistics listed in the Bibliography. For our present purpose we shall restrict ourselves to a brief description of the problems appearing at the various levels of complexity and uncertainty.

(a) Difference-Type Experimenting in the Physical Sciences

Suppose, for example, we wished to study some relatively small effect, such as the extension of a hard steel wire under load. Not only is the effect small, but it also is subject to a number of perturbing factors—for example, temperature. If we simply measure, therefore, the extension under a certain load without ensuring temperature stability, we cannot be certain that the extension we measure can be ascribed uniquely to the influence in which we are interested, namely load. Not only that, but if, in addition, we are actually unable to control the temperature, we shall never be able to be sure about the effect of load on the wire. The solution is a "null-effect" measurement. If we simultaneously measure the length of two specimens, one loaded and the other not, we may hope to ascribe the difference in their behavior to the quantity in which we are interested, the load. We must obviously try to ensure that, as far as possible, the two specimens be identical, be subject to exactly the same influences, such as temperature, and differ only in the one respect, load.

Fortunately such correspondence is not too hard to achieve if we are talking about steel wires. We can come close to making the situation of the two wires identical by mounting them close together (to minimize temperature differences between them) and by taking other similar precautions. And, since we wish the basic properties of the two specimens to be as close to identical as possible, we can simply take one length of wire and cut it in two, making one piece the specimen to be loaded and the other the comparison specimen that will indicate the null effect. Our ability to cut the specimen in half allows us conveniently to perform a great variety of difference-type measurements and to obtain high precision in the detection of small effects that would otherwise be hopelessly obscured by perturbing factors. Such difference-type experimenting is very common and will be encountered over the whole range of physical phenomena.

We are very frequently led into error unless we check the performance of our experimental system *in the absence* of the influence we are studying as well as in its presence. Sometimes the results surprise us, and we would do well to take the advice of Wilson (see the Bibliography) and reflect on the statement: "It has been conclusively proved by numerous tests that the beating of drums and gongs during a solar eclipse will cause the sun's brightness to return."

(b) Difference-Type Experimenting in the Biological Sciences

In illustrating null-effect measurements using the extension of a loaded steel wire we have encountered one very convenient aspect. In order to guarantee the similarity of the experimental specimen and the comparison specimen we took our basic specimen and cut it in half. In the case of steel wires and other similar materials, that presents no problem. Other systems, however, are not so cooperative.

Suppose we wished to measure the effectiveness of a new drug for a particular type of illness. It would clearly be of little value if we did nothing other than simply administer the drug to a patient suffering from the disease and watch for improvement. There are far too many variable and perturbing factors for us to ascribe confidently any change in the patient's condition to the drug. If we wish to isolate the effect of the drug alone, we should clearly try to design some sort of difference-type experiment in which we observe the null effect as well as the influence of the drug. Such a requirement, however, raises obvious difficulties not encountered when experimenting on steel wires. The reluctance of most human specimens to be cut in half makes it impossible to create a genuine null-effect specimen. We could use a second person as a null-effect specimen, but we would immediately encounter all the variability of response that we had sought to evade by using identical specimens.

Faced with the inevitability of biological variability, our only recourse is to compensate with increased numbers. We abandon attempts to experiment on single specimens and construct an experimental group, exposed to the influence under study, and a "control" group. The control group is constructed to be as closely comparable as possible to the experimental group, differing only in that it will not receive the treatment that forms the topic of the research. It will, we hope, be exposed to all the perturbing influences that affect the experimental group, will respond to them in the same way as that group, and will, therefore, provide the null-effect measurement.

Many refinements may have to be built into this kind of experimenting, because the effects we seek to measure can often be quite small in comparison with all the perturbing influences. For example, to diminish subconscious distortion of the results in medical experimenting on human subjects, it is common to offer the members of a control group a simulation of the real material given to the experimental group (a "placebo"), while keeping both the experimenters and the subjects in ignorance of the allocation of real and simulated material (the so-called "double-blind" experiment).

Experiment designs involving an experimental group and a carefully matched control group are virtually universal in biological studies, whether we are trying to measure the possible carcenogenicity of some food dye in large numbers of unfortunate mice, or the beneficial effects of musical activities on the academic achievement of elementary-school students.

5-7 EXPERIMENTING WITH NO CONTROL OVER INPUT VARIABLES

Sometimes we have to design a process to study some system over which we have no control at all. If this is the case, we have no alternative to simple, unmanipulative observation of the system, and our task is to design the observational procedure (perhaps we are not justified in calling it an experiment) so as to optimize our chances of effective comparison between the properties of the system and those of any model we have in mind. In cases of clear-cut behavior of the system and well-defined models we may not have too much of a problem. For example, an astronomer may suffer the frustration of inability to influence his subject matter, but his system usually functions in a well-defined manner, often permitting extremely accurate measurement. In this way it is not too hard to decide that Einstein's theory of general relativity fits the observations on the orbit of the planet Mercury better than does Newton's theory of gravitation.

In other cases, however, the questions we ask may be harder to answer.

For example, has the introduction of a new detail of manufacture altered the quality of a manufactured product or not? Even when everything in the manufacturing process is being kept as nearly constant as possible, observation will show that the product varies from specimen to specimen; does this variance mask the effect in which we are interested or not? In such cases, without control over our input variables, our study becomes an exercise in sampling procedures, and a whole field of industrial study exists under the title of quality control. The literature on statistics and statistical experiment design is very extensive, but some of the texts listed in the Bibliography will provide a starting point.

Even industrial processes, however, with their inherent fluctuations and their lack of input control pose problems that are simple in comparison with some of the questions to which we seek answers today. For example: Does the addition of fluorides to municipal water supplies improve the condition of people's teeth and does it have other, possibly harmful, effects? Do nuclear power stations cause a higher incidence of leukemia in their vicinity or not? In such cases we have almost every problem that can face an experimenter. There is little or no control over the input variables, there is wide variation in individual response, the response may be only of a probabilistic nature, there may be long delays in observing a response, there is rarely an opportunity to observe a genuine null effect (we do not normally carry out surveys of sufficient sensitivity before the municipality starts to add fluorides or before the nuclear power station is built), and there is commonly a multitude of confusing extraneous factors. The only thing we can do is to carry out our sampling procedure as carefully as possible. We must obtain an artificial null-effect measurement by constructing an experimental group, as large as possible, that will be under the influence we are studying and a control group, hopefully exempted from that influence, that in every other respect matches the experimental group as closely as possible.

The whole point in this kind of experimenting or survey work lies in the skill and care with which the sampling is done. The effects under study are usually so subtle that, as a consequence of no more than changes in sampling procedure, it is not uncommon for different surveys to provide completely contradictory conclusions. In fact it is not completely unknown that people with special interests in mind can supply results of surveys to "prove" their point, obtaining the result they want by careful control over their sampling procedures. Many of the issues in which scientific matters have a bearing on public policy have this characteristic of uncontrollable input variables, and we should all become as familiar as possible with the procedures used for sampling and significance testing. In this way we shall be able to judge as accurately as possible the usually conflicting claims of the protagonists.

When faced with problems of such complexity, we must frequently abandon familiar patterns of thought that have been successfully used in other areas. For example, the word "proof" is legitimately used in many contexts. We can prove mathematical results as consequences of mathematical principles. The word is also used (perhaps less legitimately) with reference to measurements when the uncertainty level permits. We have proved that the sun is more distant from the earth than the moon (although it would be better to say simply that the distance is measured to be greater). But there are other areas in which we cannot use the word at all. We have all heard, for example, that the evidence linking cigarette smoking to lung cancer is "only statistical" and that harm has not been "proved." Such situations are confusing, partly because the observable effects appear only in terms of probabilities, and also because the effects may appear only after years have elapsed. In such cases the concept of "proof" must be modified. It is actually replaced by the concept of *correlation*. Correlation studies give results, phrased in terms of probabilities, that differ in character from the clear-cut cause-and-effect relationships with which we are familiar in other experiments. Nevertheless they can be equally valid for identifying the factors that influence systems. The concept of correlation will receive further consideration in Sec. 6–14.

PROBLEMS

1. A scientist claims that the terminal velocity of fall of a parachutist is dependent only on the mass of the parachutist and the acceleration due to gravity. Is it worth while setting up an experiment to check this?

2. The range of a projectile fired with velocity v at angle α to the horizontal may depend on its mass, the velocity, the angle, and the gravitational acceleration. Find the form of the function.

3. The pressure inside a soap bubble is known to depend on the surface tension of the material and the radius of the bubble. What is the nature of the dependence?

4. The period of a torsion pendulum is a function of the rigidity constant (torque/unit angular deflection) of the support and the moment of inertia of the oscillating body. What is the form of the function?

5. The deflection of a beam of circular cross section supported at the ends and loaded in the middle is dependent on the loading force, the length between the supports, the radius of the beam, and Young's modulus of the material. Deduce the nature of the dependence.

In all the following problems state the variables or combination of variables which should be plotted to check the suggested variation and state how the unknown may be found (slope, intercept, etc.).

6. The position of a body starting from rest and subject to a uniform acceleration is

described by

$$s = \tfrac{1}{2}at^2$$

s and t are measured variables. Determine a.

7. The fundamental frequency of vibration of a string is given by

$$n = \frac{1}{2l}\sqrt{\frac{T}{m}}$$

n, l, and T are measured variables. Determine m.

8. The velocity of outflow of an ideal fluid from a hole in the side of a tank is given by

$$v = \sqrt{\frac{2P}{\rho}}$$

v and P are measured variables. Determine ρ.

9. A conical pendulum has a period given by

$$T = 2\pi\sqrt{\frac{l\cos\alpha}{g}}$$

T and α are measured variables, l is fixed and known. Determine g.

10. The deflection of a cantilever beam is expressed by

$$d = \frac{4Wl^3}{Yab^3}$$

d, W, and l are measured variables, a and b are fixed and known. Determine Y.

11. The capillary rise of a fluid in a tube is given by

$$h = \frac{2\sigma}{\rho g R}$$

h and R are measured variables, ρ and g are fixed and known. Determine σ.

12. The gas law for an ideal gas is

$$pv = RT$$

p and T are measured variables, v is fixed and known. Determine R.

13. The Doppler shift of frequency for a moving source is given by

$$f = f_0\frac{v}{v - v_0}$$

f and v_0 are measured variables, f_0 is fixed and known. Determine v.

14. The linear expansion of a solid is described by

$$l = l_0(1 + \alpha \cdot \Delta t)$$

l and Δt are measured variables, l_0 is constant but unknown. Determine α.

15. The refraction equation is

$$\mu_1 \sin \theta_1 = \mu_2 \sin \theta_2$$

θ_1 and θ_2 are measured variables; μ_1 is fixed and known. Determine μ_2.

16. The thin-lens (or mirror) equation can be written

$$\frac{1}{s} + \frac{1}{s'} = \frac{1}{f}$$

s and s' are measured variables. Determine f. There are two ways of plotting this function. Which is the better?

17. The resonant frequency of a parallel L-C circuit is given by

$$\omega = \frac{1}{\sqrt{LC}}$$

ω and C are measured variables. Determine L.

18. The force between electrostatic charges is described by

$$F = \frac{q_1 q_2}{4\pi\epsilon_0 r^2}$$

F and r are measured variables for fixed and known q_1, q_2. How do you check the inverse-square law?

19. The force between currents is described by

$$F = \frac{\mu_0}{2\pi} \frac{i_1 i_2 l}{r}$$

F, i_1, i_2, and r are measured variables, μ_0 and l are constant. How do you check the form of the dependence?

20. The discharge of a capacitor is described by

$$Q = Q_0 e^{-t/RC}$$

Q and t are measured variables. R is fixed and known. Determine C.

21. The impedance of a series R-C circuit is

$$Z = \sqrt{R^2 + \frac{1}{\omega^2 C^2}}$$

Z and ω are measured variables. Determine R and C.

22. The relativistic variation of mass with velocity is

$$m = \frac{m_0}{\sqrt{1 - \frac{v^2}{c^2}}}$$

m and v are measured variables. Determine m_0 and c.

23. The wavelengths of the lines in the Balmer series of the hydrogen spectrum are given by

$$\frac{1}{\lambda} = R\left(\frac{1}{4} - \frac{1}{n^2}\right)$$

λ and n are measured variables. Determine R.

6

Experiment Evaluation

6–1 GENERAL APPROACH

Even when we have finished making the measurements in an experiment, an equally significant part of the process still remains; we must evaluate the significance in what has been done. Our objective in doing an experiment is to be able to make some statement. It is important to identify clearly the statement we wish to make and to ensure that the statement is as accurate and complete as possible and fully justified by our measurements. The precise way in which we evaluate the experiment as a whole depends on the type of experimenting we have been doing. As described in Chaps. 4 and 5, we may have been operating with or without a theoretical model, and our measurements may or may not have been dominated by statistical variance. The procedures we must now follow will vary accordingly.

Before we proceed, however, we must note two general points. First, we should always remember that experimental results are precious. Often they have been obtained from an extensive experimental program involving many people and large amounts of money. At any level of cost the results may be unique and irretrievable. We should accept the obligation to extract every available bit of information from our observations and to ensure that our final statement is as complete as possible. The second general remark concerns objectivity. It is almost impossible to avoid approaching an experiment without some preconception of what "ought" to happen. We must, however, discipline ourselves to be as objective as we can, and if the outcome of the experiment is different from what we expected or hoped for or is disappointing in some way, we must be prepared to state the result honestly and realistically and to obtain from it the guidance required for future work.

In the teaching laboratory, where it is sometimes difficult to keep our ultimate objectives clearly in mind and easy to forget that our experiments serve to simulate real tasks in the working world, we commonly encounter the mistaken belief that the task is to reproduce the known values of experimental quantities. If we are measuring the acceleration of gravity and obtain a value of 9.60 m sec^{-2}, our answer is different from the "right" answer and so we are "wrong." The "error" can then be conveniently blamed on the apparatus. In fact, since there are no "right" answers for experimental quantities, our situation really involves the comparison of two measured values for a quantity. Each measured value has its own characteristics, and each has its own range of uncertainty. To assess the significance of a discrepancy between two independently measured values of a quantity is actually a complex and difficult task. It is far better at first to develop our ability to make measurements as reliably as possible and to assess their range of uncertainty as accurately as possible; we can worry later about comparing our measurements with those of other people.

So when we are making measurements on quantities for which we are sure we already know a more precise value, it is best to discipline ourselves to avoid thinking about the more precise or "standard" value; it is better to acquire experience and build up confidence in our own work. This confidence will be necessary later as we undertake professional experimenting, in which we must take responsibility for our experiments and measure things that have never been measured before.

So if we obtain 9.60 m sec^{-2} for g, let us be equally aware that our measured uncertainty is ±0.3 m sec^{-2} and our result is not as bad as we might think at first. If we are going to grumble about anything, let it be the ±0.3 m sec^{-2}, but we must not feel guilty about it if the experimental apparatus, with normal effort, is not capable of precision better than 3%. We must not be misled by the way in which accepted values for physical quantities are quoted in textbooks. The values are often mentioned rather casually, and the texts rarely make it clear that these numbers represent the outcome of sophisticated work by generations of expert scientists. It is instructive for us to read the detailed history of such measurements, and excellent accounts of some of them will be found in the book by Shamos listed in the Bibliography. We should not be too casual about numbers such as these and should not hope to reproduce them exactly in two hours of work in an elementary laboratory.

The main point is to state the result of the experiment honestly and objectively. Certainly the experimenter should strive earnestly to maximize the yield of the experiment by making the final answer as reliable as possible and the limits of uncertainty as close as the experiment will permit, but in all cases it is important to be realistic.

6–2 THE STAGES OF EXPERIMENT EVALUATION

The process of evaluating the result of an experiment has several parts. First we must obtain the values of the basic measurements and their uncertainties. Second we must assess the degree of correspondence between the properties of the system and of the model. Third we must calculate the values of whatever property of the system we had set out to measure. Last we must make an estimate of the overall precision of the experiment. Let us consider each of these steps in turn.

(a) Computation of Elementary Quantities

The first step in working out the result of our experiment consists of calculating the elementary quantities of which the experiment is composed. For example, an experiment on a simple pendulum that has the purpose of obtaining a value for g will probably yield, as its input variable, a set of measurements of length ℓ. The output variable will be presented by a set of measurements of the times required for a certain number of oscillations, and from them values of the period T can be calculated. Our present purpose is to compute the values of ℓ and T and their uncertainties; these will be the basis of the subsequent graphical analysis. The choice of procedure here will depend on whether we have elected to make a subjective assessment of the uncertainty range of each measurement or have decided that random fluctuation is sufficiently prominent that statistical treatment is desirable.

(b) Estimated Uncertainty

In the case of the simple pendulum the first variable to consider is ℓ. Here we may have found that measuring the length of the pendulum with a meter stick has enabled us to identify intervals, as described in Sec. 2–3, within which we are "almost certain" our values lie. Our experimental results will therefore take the form of a set of values for ℓ in the form: value ± uncertainty. It is conceivable, too, if we have been counting swings and measuring the times with a stopwatch, that we are similarly able to identify intervals on the time scale within which we are "almost certain" that our values for time lie. These, too, would be expressed as time value ± uncertainty. This, however, is not yet our variable T. We might have counted 15 oscillations of the pendulum, obtaining a time value of 18.4 ± 0.2 sec, and the value for the period, the time required for *one* oscillation, must be obtained by division as 1.227 ± 0.013 sec. Note that not only the central value must be calculated in this way but also the value of the uncertainty. In simple, algebraic terms

$$\left(\frac{1}{15}\right)(18.4 \pm 0.2) = \frac{18.4}{15} \pm \frac{0.2}{15} = 1.227 \pm 0.013$$

Do not ignore this kind of significant modification of the uncertainty value; it will be necessary whenever any arithmetic process is carried out on the basic measurements.

The end result of the process for this experiment will be a set of ℓ and T values, complete with uncertainties, and we would be ready to consider drawing our graph.

(c) Statistical Uncertainty

If repetition of our measuring process has shown random fluctuation in one or both of the variables, we may have decided, as described in Sec. 5–3(e), to take a sample of readings, the number of readings being chosen on the basis of the apparent magnitude of the scatter to give the precision we require. Since we must reduce the resulting set of readings to a form suitable for plotting, we must express our sample in the form: central value ± uncertainty. As described in Sec. 3–10, the most suitable form to choose is usually the sample mean and the standard deviation of the mean because of the readily identifiable significance of these quantities. Provided we make it clear in our report that we are quoting sample means and standard deviations of means, everyone will understand that we are specifying intervals that have a 68% chance of containing the universe mean.

While we are making these claims about the numerical significance of our measurements, it is worth remembering the warnings given in Sec. 3–5. The measurement samples encountered in the work of the physics laboratory are frequently too small to permit any assessment of the actual frequency distribution of the universe from which the measurements were taken. We are therefore making an assumption when we ascribe the numerical properties of the Gaussian distribution to our sample. It is usually a good enough assumption, but we should remember clearly that it is an assumption.

At this point remember also the warnings about σ estimates from small samples that were given in Sec. 3–11, and check that the computations are significant. In general it is not worth using a statistical approach with fewer than 10 observations; for particular purposes many more may be required.

It is useful to think in advance about the interpretation of the uncertainty regions on the graph. If both variables in our experiment have similar statistical character, the mean and standard deviation of the mean for each point will enable us to draw, for each point on the graph, a little rectangle whose interpretation will be clear. We may have a little more of a problem if our experiment has yielded variables of two different kinds. It is quite conceivable in, say, the experiment on free fall under gravity used as an example in Sec. 4–2, that one variable, the distance of fall, will appropriately have an estimated uncertainty and the other will require a statistical treatment yielding standard deviations of the mean. If we were to plot values derived from these two different types of treatment, our uncertainty ranges along the two axes would be different. The uncertainty interval in one dimension would give almost 100% probability of containing our desired value, while the probability in the other dimension would be only 68%. In a case such as this, it would be difficult to know how to interpret the graph, and it would be better to bring the two variables into better correspondence. Remembering that a range of twice the standard deviation of the mean gives us a 95% chance of including the universe value, we can use $2S_m$ as our uncertainty for the statistically treated variable, thus giving a range of uncertainty for each point on the graph with roughly the same significance in both dimensions.

At this stage, by one process or another, the measurement of every quantity in the experiment will have been reduced to a central value and its uncertainty, but we are not quite ready to start drawing the actual graph yet. If the graph is to be drawn with one variable on one axis and the other variable on the second axis (like load vs. extension for a spring or current vs. potential difference for a resistor), then we can proceed directly. However, it is equally common to plot quantities on the graph that must be constructed out of the elementary measurements by some process of arithmetic compution (T^2 vs. ℓ for a pendulum, t vs. $\sqrt{h}$ for free fall under gravity, etc). There is obviously no problem in performing such simple arithmetic calculations, but we must not forget that the uncertainty values also must be recalculated. If we are going to plot values of T^2 on the graph, the uncertainty bars or rectangles must give the actual interval over which T^2 itself is uncertain. All such computed quantities must be provided with their own uncertainty intervals, and only then are we ready to start drawing the graph.

6–3 GRAPHS

Whether the graph is to be merely an illustration of the behavior of a physical system or whether it is to be the key to assessing the experiment and calculating the answer, the aim is to set out the results in such a way that their characteristics are displayed as clearly as possible. This will involve appropriate choices of scale, proportions, etc. First, ensure that the graph paper is large enough. It is a waste of time to plot observations having a precision of $\frac{1}{5}$% on a piece of graph paper 12 cm × 18 cm, where a typical plotting uncertainty is perhaps 2%. As we shall see later, valuable information will be lost unless the uncertainties on the points are clearly visible, and so it is necessary to make sure that the graph paper is big enough. Second, make the graph fill the available area. This can be done by choosing the scales so that the general course of the graph runs at about 45° to the axes and by suppressing the zero if necessary. If one is plotting the resistance of a copper wire as a function of temperature and the values run from 57 to 62 ohms, start the resistance scale at 55 ohms and run it to 65. If the scale is started at zero, the graph will look like a flat roof over a sheet of empty graph paper and convey very little information.

There are exceptions, however, when it may be important to preserve the origin as part of the graph. It may be desirable, or even necessary, to examine the behavior of the graph at the zero of one or both axes. At other times, for purposes of illustration, we may wish to show clearly the scale of some variation in relation to its zero value. However, for the purposes of the graphical analysis with which we are here concerned, it is generally best to make the graph fill the graph paper.

The method of marking each measurement on the graph paper depends to some extent on preference. One essential feature, of course, is to make sure that the range of uncertainty is clearly indicated. Only if this is done can the process of comparing the behavior of the system and the model have any meaning and the uncertainty of any future calculations of slopes, etc., be assessed. To each point on the graph we can attach a cross, with horizontal and vertical bars to indicate the range of uncertainty, or we can make each "point" a little rectangle surrounding the measured value and indicating by its horizontal and vertical dimensions the range of uncertainty in each coordinate. So long as the ranges of uncertainty are clearly indicated, it may not matter which method we choose; the important thing is to acquire the habit of marking uncertainties on every graph we draw. It is also important to note on the graph itself, or in its caption, the nature of the uncertainties—i.e., estimated outer limits of uncertainty, statistical uncertainties of $1S$ or $2S$, etc. It can be very frustrating when trying to judge the significance of a graph if we have to search through the text to find out what the uncertainty marks mean. If there are several graphs to be plotted on one piece of paper, make sure they are clearly distinguished by the use of some different symbol or by color or by some other means.

6–4 COMPARISON BETWEEN EXISTING MODELS AND SYSTEMS

Once all our observations are plotted on the graph paper, we are ready to proceed with the next stage—the comparison between the properties of the system, now displayed before us, and the properties of any models we have available. Our procedure will depend on circumstances, and we shall describe the various situations in turn. In all of the following we shall assume that, on account of the difficulty in representing nonlinear properties of models on hand-drawn graphs, we have chosen or rearranged our variables so that the graphs will take on linear form.

Let us suppose, firstly, that we have a model that is fully specified and contains no undetermined quantities. The purpose in our experimental investigation would then be only to see how well the properties of the model match the properties of the system. To do this we would simply draw on our graph, using the same scale, the graph of the function that represents the properties of the model. A typical case was illustrated in Fig. 4–10, in which observations of the time of fall of an object as a function of distance are compared with the behavior of the analytical expression

$$t = 0.4515x^{1/2}$$

which represents our theoretical model of the situation.

But how are we to judge the degree of correspondence? This is where the presence of the uncertainty ranges becomes of dominating importance. If we had simply plotted points without uncertainty bars, the inevitable scatter in the points would mean that the probability that the line representing the model's properties actually passes through even one (not to say more than one) of the points would be vanishingly small. So how would we be able to say anything sensible about the outcome of the comparison? If, however, the "points" on the graph represent ranges of possibility for the location of the plotted values, it becomes possible to make logically satisfactory statements. If, as was the case in Fig. 4–10, the line representing the model passed through the region of uncertainty of each point, we could say just that. Note once again that this does not mean that we have "proved" that the equation is "true," or "correct," or whatever. All we can say is that the model and the system are "consistent," or "in agreement," or "compatible," or some such phrase. We must make sure that we use the right language, for otherwise we are misrepresenting the situation and have a good chance of misleading people. Note also that we must be careful to say that we have found "correspondence," "consistency," "agreement," or whatever, between the model and the system only at the level of precision of our experiment. Nothing in our process entitles us to ignore the fact that, at a higher level of precision of measurement, discrepancies might appear that were undetectable in our experiment.

Now that we have considered the case in which a model and a system turned out to have properties that were undistinguishable at the level of precision involved, we must consider the other possibilities in which the properties of the model and the system do not overlap completely. Let us consider the various possibilities in turn.

(a) No Detectable Discrepancy

This is the case we have already considered in detail. It is illustrated in Fig. 6–1(a).

(b) Correspondence over Part of the Range

Circumstances are very frequently encountered in which a model provides a satisfactory description of a system, provided the value of some variable does not exceed or fall below some limit. In this case the graphical comparison would appear as in Fig. 6–1(b) or (c). An example of the case (b) would be the flow of a fluid through a pipe, in which the proportionality between flow rate and pressure head is satisfactory only below the onset of turbulence. Figure 6–1(c) could be a representation of the variation with temperature of the resistivity of a metal, for which the linear model breaks down at the lowest temperatures.

In any case that comes within this category we would state the result of the comparison using some phrasing such as "We observed agreement (compatibility, consistency, etc.) between the model and the observations only over the range so-and-so; the properties of the model and the system are observed to diverge significantly after the value such-and-such." Note again that we must resist the temptation to think that something is "wrong" because we do not encounter complete correspondence between models and systems over the whole range. Models and systems both exist in their own right, and we cannot prejudge the extent to which their properties overlap. In fact the detection of the limits on the validity of a particular model can furnish important clues for its improvement.

(c) Intercepts

A frequently encountered circumstance involves intercepts. The graph of the model's behavior may pass through the origin but the observed behavior of the system may not, as illustrated in Figs. 6–1(d) and (e). Such a discrepancy can

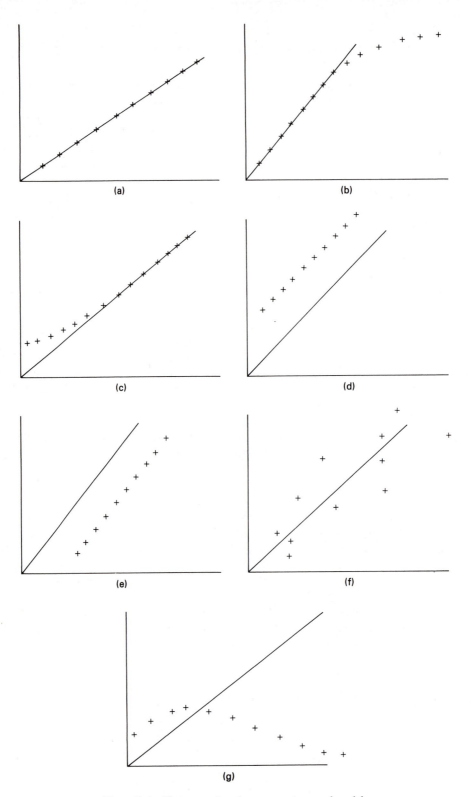

Figure 6-1 The comparison between systems and models.

arise from many different types of mismatch between the model and the system, and information about such intercepts can be very helpful in analyzing experimental situations. It is usually well worthwhile, while drawing graphs, to check the behavior of the model and system at the origin; reference was made to this point in Sec. 6–3 on graph drawing. As we saw in Sec. 4–5, the graphical analysis of an experiment is invaluable for enabling us to obtain answers that are free from the systematic errors associated with unexpected intercepts. Even with this protection, however, it is usually best to know whether an unexpected intercept exists so that we can check the overall degree of correspondence between the model and the system.

(d) Unexpected Scatter of Points

As was described in the chapter on experiment planning, we should have carefully judged the uncertainty of our measurements before starting the experiment itself, and, in the light of our target value for final precision, we should have made appropriate choices for our measurement methods. If we have done this satisfactorily, we shall find, on plotting the graph, that there is consistency between the scatter of the points and the uncertainties of the measurements. This was illustrated in Fig. 6–1(a). However, things do not always work out as we would wish, and we not uncommonly find ourselves in the situation illustrated in Fig. 6–1(f). It results, simply, from the presence of factors in our measurement methods that we failed to identify as we made our initial assessment of the uncertainty of the measurements.

We should not be content to leave the situation like this. It is worth checking the apparatus in an attempt to discover the cause of the fluctuation. It could be something as simple as a loose electrical connection or failure to stir a heating bath, and it is always satisfying to see such a discrepancy disappear. If, for any reason, it is not possible to keep the experiment going and take steps to reduce the scatter, it may be necessary to work with the results as they are and make the best statement we can about the degree of correspondence between the model and the system. We might be able to say something like "the observations are distributed uniformly about the line representing the model." For cases in which we have to obtain numerical information from lines drawn using such experimental measurements, see Sec. 6–7.

(e) No Correspondence Between System and Model

Very rarely we encounter circumstances in which the behavior of the system bears no resemblance at all to the behavior of the model [Fig. 6–1(g)]. If everything in the experiment is working as it should, this is a most unlikely outcome. Models may be, in principle, inadequate representations of the behavior of the physical world, but they would not have the status of models if they were as bad as we are currently suggesting. Such complete failure of correspondence points clearly to an actual error in the experiment. It could be an error of interpretation of the variables, a mistake in the rectification of the equation, an error in setting up the apparatus, or a mistake in making the observations, in calculating the results, or in plotting the graph. If possible, go back to the beginning, check everything, and start again. If it is not possible to check the instrumental aspects of the experiment, check for errors in all the analytical and arithmetic processes. If every attempt to discover an error fails, state the outcome of the experiment honestly and objectively. There is always the chance that we have discovered something new. In any case, if we are truly baffled by some failure of correspondence between a well-checked piece of apparatus and a reliable model, an honest statement of our situation is bound to be of interest to other people.

In all the foregoing we have been trying to stress one important point. We must not think that the experiment is giving us a "right" or a "wrong" result. We just carry out our experimental process as carefully as possible and then state the outcome as honestly and objectively as we can. It is not a bad thing for us to be reminded from time to time that models may provide only partially satisfactory representation of the behavior of systems. It is most important for us to know the limits for the validity of models, and the manner in which models fail can furnish invaluable evidence to those who seek to improve them.

6–5 CALCULATION OF VALUES
FROM STRAIGHT-LINE ANALYSIS

In all of the preceding sections we have been dealing with models that were completely specified, including the numerical values of all quantities. In such cases the experimental purpose was simply to compare the behavior of the system and of the model. As was considered in Sec. 4–5, however, it is also possible, indeed very common, to use straight-line analysis in determining, for some quantity in the model, the numerical value that is appropriate for our system. In such cases our model is not fully specified, because it contains a quantity or quantities of initially unknown value. It is not possible, therefore, to draw a graph for the model to compare with the points. In this case the graph initially contains nothing but the points alone, as shown in Fig. 6–2(a).

Let us suppose that we have been measuring the values of current through, and potential difference across, a resistor and we wish to test the observations against the model

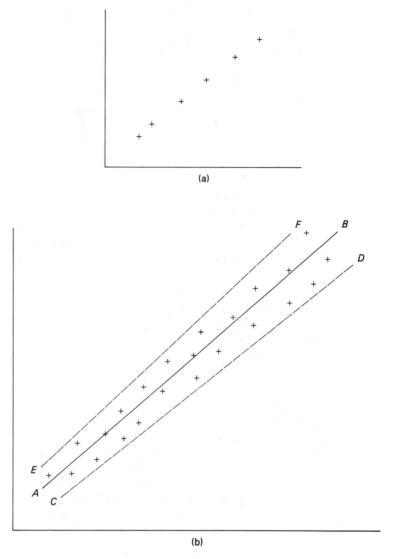

Figure 6–2   Fitting a straight line to observations.

$$V = IR$$

In the absence of a specified value for R, the behavior of the model represented by all the lines on the I, V plane that have the equation

$$V = \text{const.} \times I$$

where the constant can take all values from zero to infinity. In principle we can simply draw all these lines on the same graph as the measurements and determine, first, the extent to which the behavior of the system and of the model overlap. Second, from the bundle of lines that fall within the regions of uncertainty of all the points, we can determine the range of R values that is appropriate for our system (this was illustrated in Fig. 4–11). For our present purpose, however, it is not quite as simple as that, because, on the basis of the measured values shown in Fig. 6–2(a), we have no right to prejudge the behavior of the system at the origin. It is best to leave the question of intercepts to a later stage and simply decide on the range of straight lines that is consistent with the observations.

There are several ways of doing this. The most satisfactory, a statistical method, will be described later. In the meantime we shall content ourselves with simpler, mechanical procedures and shall carry out the time-honored practice of "drawing the best straight line through the points." To do this by eye requires some mechanical aid that does not obscure half the points. An opaque ruler cannot be used, and a transparent straight edge is acceptable. Probably the most satisfactory aid is a length of dark thread. It can be stretched over the points and easily moved until the most satisfactory position is found. If difficulty is encountered in judging visually the trend of a set of points, it is often very helpful to hold the graph paper at eye level and sight along the points. This makes clustering of the points around a straight line, or systematic deviation from a straight line, much clearer than in the direct view.

There are several straight lines that we can profitably identify. The "best" straight line (whatever we mean by "best") is one obvious candidate. In addition we can make a guess at how far we can twist that "best" line in either direction until it can no longer be regarded as an acceptable fit to the points. These two lines will supply us with a value for the uncertainty in the slope. In cases for which we find it difficult to identify a "best" line and its uncertainty limits it is sometimes helpful to remember that our points and their uncertainties are really a sample from a whole band of values on the plane. The occupation of this band by our results may be spotty because of the limited number of observations, and this makes it difficult for us to choose our lines. If this is the case, it is often helpful to imagine the band to be populated by the million or so readings that we could have made with the apparatus. We can try to guess from our graph where the center and the edges of that band might be, and that will enable us to make our choice of lines. In Fig. 6–2(b) we could have chosen AB as our "best" line and we could have decided that the lines CD and EF would contain almost all of the infinite universe of points. The lines CF and ED (not drawn) would then represent, respectively, the steepest and least steep slope of lines that are consistent with the observational points.

Once we have chosen our lines, we can set about determining the numerical value of their slopes so that we can obtain the answer we want, such as, in the case of the $V = IR$ example, the value of R. Note that, for our purpose, the question of slope has nothing to do with the angle made by the lines on the graph paper; we are talking about intervals of the measured variables I and V, and the slopes must be calculated analytically. For a line such as AB on Fig. 6–3 look carefully near the ends and identify places where the line crosses, as exactly as possible, an intersection of lines on the graph paper. Identify the coordinates (I_1, V_1) and (I_2, V_2) of these intersections and evaluate the slope as

$$\text{slope} = \frac{V_2 - V_1}{I_2 - I_1}$$

We then have, immediately,

$$R = \text{slope}$$

giving us the answer we want. In more complicated expressions, of course, the value for the slope may give our desired answer only after computation with other measured quantities.

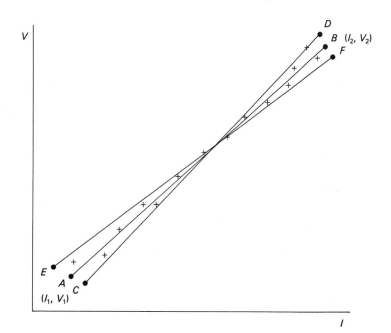

Figure 6–3   The "best" slope and outer limits.

We should carry out this process three times. Our line *AB* will give us our chosen "best" value for *R*, and the other two lines, *CF* and *ED*, will give us the upper and lower limits, outside which we are "almost certain" the value for *R* does not lie. It will usually happen that these extreme values for the uncertainty range are roughly equidistant from the central value, and it will be finally possible to state our value for *R* as

$$R = \text{value} \pm \text{uncertainty}$$

It may sometimes appear that our "best" line and the two outer limiting lines are not equally spaced. The reason is usually that the graph contains too few points to allow good assessment of the positions of the lines. Although circumstances occasionally appear in which experimenters feel obliged to express a result as

$$\text{value} \begin{cases} + \text{ uncertainty 1} \\ - \text{ uncertainty 2} \end{cases}$$

visual judgment of a graph is rarely sufficiently precise as to justify such a procedure. The positions of the lines on the graph should be reassessed in a search for a more plausible, symmetrical configuration. If, in our experiment, the desired answer is not equal to the slope directly, our expression for the slope may contain a number of quantities, and our unknown may have to be calculated from the slope by a separate arithmetic process. If these other quantities are themselves uncertain, the uncertainty in the answer will have to be found by combining the uncertainty of the slope with the other uncertainties, using the techniques of Chap. 2.

It is natural at this stage to think about the significance of the uncertainty associated with quantities obtained from graphs. This will depend on the type of uncertainty marked on the graph. If the bars indicate outer limits of possible variation (subjectively assessed or $2S_m$ in the case of statistical fluctuation), then the limits on the slope will similarly be of this nature. If the points have been marked with $1S_m$ limits, the limiting slopes *CF* and *ED* will probably represent limits implying better than 68% probability, since the limiting lines are drawn with a pessimistic bias.

We have assumed in the foregoing that the scatter encountered in the actual results is within the predicted range of uncertainty. Then the use of the limiting lines gives rise to a fairly well-defined value for the uncertainty in slope. If, however, the scatter is well outside the expected range of uncertainty (owing to an unsuspected source of fluctuation), then there is no unique setting for lines within which we are "almost certain" the answer lies. In such a case, and in all precise work, there is no substitute for the method of least squares, which will be described in Sec. 6–7.

Note that in choosing our three lines we have deliberately excluded the origin from consideration. We did this precisely because the behavior of the system at the origin may be one of the things we wish to examine. If the graph of the model's behavior does pass through the origin, we should inspect our three lines in that region. It is most unlikely that our central line will pass exactly through the origin, but, if the area between the two limiting lines does include the origin, we can say that we have consistency between the model and the system, at least at our present level of precision. Only if both limiting lines clearly intersect an axis on one side of the origin can we claim that we have unambiguously identified an unexpected intercept.

If the behavior of the model does lead us to expect an intercept from which we hope to obtain the value of some quantity, the intersection of the three lines on the axis in question will give us that intercept directly in the desired form, value ± uncertainty.

6–6 CASES OF IMPERFECT CORRESPONDENCE BETWEEN SYSTEM AND MODEL

In cases where the correspondence between model and system is only partial, we must be careful to obtain our answers without introducing systematic error from the discrepancies. Referring to Figs. 6–1(b) and (c), consider first the cases in which the measured values correspond adequately with the straight line of the model only over a limited range. Obviously our evaluation of slopes should be confined to those regions in which the system and model are compatible. The points that deviate systematically from the line clearly arise from physical circumstances not included in the model, and it is obviously inappropriate to include them in any calculations based on the model. We disregard, therefore, all points that deviate systematically from straight-line behavior by an amount clearly in excess of the estimated uncertainties and observed scatter of the points, and we obtain our slope and its uncertainty from the linear region.

A second point to note concerns intercepts. Even if the model's behavior passes through the origin, it is not uncommon to find that our graph shows an intercept. Such a deviation can arise from a variety of causes; fortunately, many of these prove to be harmless. If the discrepancy causing the intercept affects all the readings in the same way (like an undetected zero error in an instrument or a spurious, and constant, emf in an electrical circuit), then the graph will give us a slope that will be free of the systematic error that would otherwise be introduced. It is wise, therefore, to arrange our experiment so that the answer will be obtainable from the slope of the graph, while quantities that may be subject to undetermined systematic error should be relegated to the role of intercepts. This capacity of graphical analysis to provide answers free from many types of systematic error is one of its chief advantages.

6–7 THE PRINCIPLE OF LEAST SQUARES

All of the procedures described in the preceding sections have one characteristic in common: they are all based on the use of visual judgment by the experimenter. Thus, although the procedures are very commonly used and are very useful, they are vulnerable to the criticism that, even when they are carefully carried out, we cannot be sure of the numerical significance of the results. It would be very comforting if we could use some mathematical procedure to identify the "best" line for a set of points, for then we would be released from the insecurity of personal judgment. In addition we could hope to find out what we mean by "best" and to assess the precision of that choice.

The procedure in question is based on the statistical principle of least squares. We shall discuss it in its restricted application to choosing a straight-line fit to measured values. (Further detail on the principle in general will be found in Appendix 2.) Consider that we have a set of N values of a variable y measured as a function of variable x. We must restrict ourselves to the special case in which all the uncertainty is confined to the y dimension; i.e., the x values are known exactly, or, at least, so much more precisely than the y values that the uncertainty in the x dimension can be neglected. If this condition cannot be satisfied, the simple treatment given below is not valid. The method can be extended to cover the case of uncertainty in both dimensions, but the procedure is not simple; the reader who wishes to pursue the subject will find an excellent treatment in the text by Wilson listed in the Bibliography.

The question now to be answered by our mathematical procedure is: which of all the lines on the x-y plane do we choose as the best line, and what do we mean by "best"? The principle of least squares makes this choice on the basis of the deviations of the points in a vertical direction from the lines. Let AB in Fig. 6–4 be one candidate for the status of "best" line. Consider all the vertical intervals between the points and the line, of which $P_2 O_2$ is typical. We shall define the best line to be the one that makes the sum of the squares of deviations like $P_2 O_2$ a minimum.

Notice that we have no right to consider an invented criterion like this to provide any automatic path to "truth" or "correct" answers. It is simply one choice of a criterion for optimizing the path of our line through the points. It does, however, offer some advantages over other possibilities, such as mini-

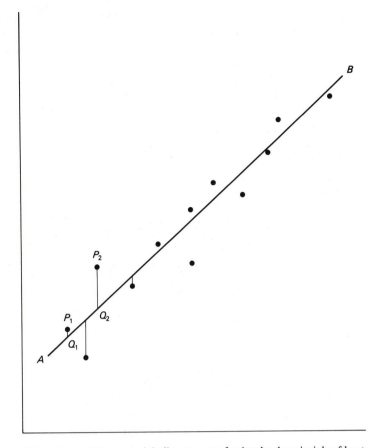

Figure 6–4   Fitting a straight line to a set of points by the principle of least squares.

mizing the third power of the intervals or the first power, etc. Although we need not, in general, concern ourselves with the rationale for the principle of least squares as we use it, it is of interest to note the basis for its claim to validity. It can be proved that the procedure of minimizing the *squares* of the deviations gives rise, on repeated sampling, to smaller variance of the resulting parameters, such as slope, than does the use of any other criterion. We are consequently entitled to greater confidence in results obtained using the principle of least squares than is the case for any competitor; consequently the use of this principle is widespread.

Let us put the least-squares principle into mathematical form. We define the best line to be that which leads to the minimum value of the sum

$$\sum (P_i O_i)^2$$

and we desire the parameters, slope m and intercept b, of that best line.

Let the equation of the best line be

$$y = mx + b$$

The magnitude of the deviation $P_i Q_i$ is the interval between a certain measured value y_i and the y value of the point, on the line, at that value of x. This y value can be calculated from the corresponding x value as $mx_i + b$, so that, if we call each difference δy_i, we have

$$\delta y_i = y_i - (mx_i + b) \tag{6-1}$$

The criterion of least squares enables us to obtain the desired values of m and b from the condition

$$\sum [y_i - (mx_i + b)]^2 = \text{minimum}$$

Write

$$\sum [y_i - (mx_i + b)]^2 = M$$

Then the condition for the minimum is

$$\frac{\partial M}{\partial m} = 0 \quad \text{and} \quad \frac{\partial M}{\partial b} = 0 \tag{6-2}$$

A brief algebraic exercise then allows us to obtain the values of slope and intercept for the best line as

$$m = \frac{N \sum (x_i y_i) - \sum x_i \sum y_i}{N \sum x_i^2 - (\sum x_i)^2} \tag{6-3}$$

$$b = \frac{\sum x_i^2 \sum y_i - \sum x_i \sum (x_i y_i)}{N \sum x_i^2 - (\sum x_i)^2} \tag{6-4}$$

We have now succeeded in replacing the sometimes questionable use of personal judgment by a mathematical procedure, giving results of well-identified significance and universal acceptability. In addition, since there is some statistical meaning in the new method, we can expect a more precise form of uncertainty calculation. In fact the least-squares principle allows us immediately to obtain values for the standard deviation of the slope and the intercept, giving us uncertainties of known statistical significance.

The standard deviations of the slope and intercept are calculated in terms of the standard deviation of the distribution of δy values about the best line, which we call S_y. It is given by

$$S_y = \sqrt{\frac{\Sigma \, (\delta y_i)^2}{N - 2}} \qquad (6\text{–}5)$$

(Do not worry about a standard deviation being calculated with $N - 2$ in the denominator rather than $N - 1$ or N; it is a consequence of applying the definition of the standard deviation to the positioning of a line on a plane.) The values of S_m and S_b are then given by

$$S_m = S_y \times \sqrt{\frac{N}{N \, \Sigma \, x_i^2 - (\Sigma \, x_i)^2}} \qquad (6\text{–}6)$$

$$S_b = S_y \times \sqrt{\frac{\Sigma \, x_i^2}{N \, \Sigma \, x_i^2 - (\Sigma \, x_i)^2}} \qquad (6\text{–}7)$$

These can be used in association with the values of m and b to indicate intervals with the normal meaning, namely that intervals of one standard deviation give us a 68% chance of enclosing the universe value, two standard deviations 95%, etc. One very important advantage of the least-squares method is that it supplies us with statistically significant values for the uncertainties in our slope and intercept that are derived objectively from the actual scatter in the points themselves, irrespective of any optimistic claims we might wish to offer for the uncertainties of the measured values.

A more complete mathematical description of the least-squares method will be found in Appendix 2. In that appendix will also be found an extension to the method that we have excluded from the present discussion. If, in our experiment, the points used in the least-squares calculation are not equally precise, we should use some procedure that accords greater importance to the more precise measurements. This procedure is called "weighting." The use of weighting procedures is not limited to straight-line fitting; they are applicable whenever we wish to combine observations in some way, even in such a simple process as determining the mean of a set of values of unequal precision. The equations for finding a weighted mean and for doing a weighted least-squares calculation will both be found in Appendix 2.

6–8 LEAST-SQUARES FIT TO NONLINEAR FUNCTIONS

The procedures used in Sec. 6–7 to determine the slope and intercept of the best straight line can obviously, in principle at least, be applied also to nonlinear functions. We can write an equation analogous to Eq. (6–1) for any function, and we can still use a requirement similar to Eq. (6–2) to express the minimizing of the quantity M with respect to the parameters in our chosen model. If the resulting equations for the parameters are easy to solve, we can proceed to obtain their values just as we did for straight lines.

Frequently, however, it is not easy to solve the equations. In such cases we abandon the attempt to obtain an analytical solution to the problem and rely on the computer to provide us with approximate solutions using iterative techniques. We construct a trial function, calculate the sum of the squared differences, and then vary the chosen function until a minimum is found for that sum. Descriptions of such computer-based methods will be found in the text by Draper and Smith that is listed in the Bibliography. If, however, a method can be found to test a model in linear form, this will certainly be simpler.

Note that it is, in all cases, the responsibility of the experimenter to choose the type of function to be used; all that the least-squares method can do is to give us, for our chosen function, those values of the parameters that provide the best fit with the observations.

6–9 PRECAUTIONS WITH LEAST-SQUARES FITTING

The mathematical procedures for least-squares fitting are completely impartial. As we use Eqs. (6–3) and (6–4) for linear fitting, they will drive a straight line through any set of points with complete disregard for the appropriateness of a straight-line function. If, for example, our experiment has given us a set of observations (Fig. 6–5) that clearly show the breakdown of a linear model and we heedlessly use the least-squares procedure on the whole set of observations, we shall obtain the parameters of a line, AB, that has no significance at all, either for the model or the system. Unthinking use of the least-squares procedures must therefore be studiously avoided.

This warning is all the more important in a day when we can all carry in our pockets little machines that can give us, at the touch of a few buttons, least-squares parameters for any set of numbers we care to insert. We must remember that, if we are comparing straight lines with our set of observations, it is because *we* have made the decision that this is a reasonable thing to do. We must not, therefore, even contemplate using a least-squares procedure until we have plotted the observations on a graph and satisfied ourselves, by visual inspection and personal judgment, that linear fitting is appropriate. In addition, it may be necessary to decide that some of the observations fall outside the scope of the model and are not appropriate for inclusion in the choice of the best straight line. Only after we have carefully considered the situation graphically

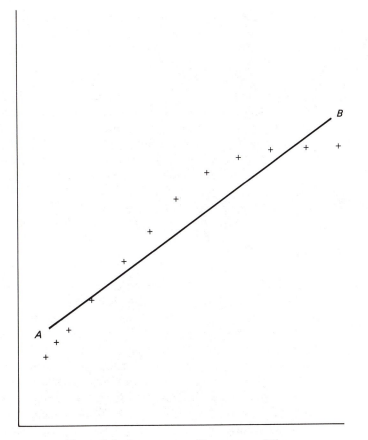

Figure 6-5   Improper use of least-squares fitting.

and assured ourselves that linear fitting is appropriate over all or part of the range of the observations are we justified in starting the least-squares procedure. Failure to heed this warning can give rise to serious error in experiment interpretation.

6-10 FUNCTION FINDING

All of the foregoing discussion involved the assumption that we were already in possession of a model which we wished to compare with a system. Although this is commonly the case, it does happen sometimes that we have a set of observations for which no model is available. This can happen, for example, in research on a phenomenon that has never been observed before, or else in our work on a system that is so complex that a theoretical model will never be available. The observations, when plotted in elementary form, will probably show a curve of no readily identifiable form. In the absence of a model, what are we to do?

One thing we can do is to try to find functions that have some degree of correspondence with the observations. Such a procedure can be very useful. For example, in very complex systems for which there is little hope of constructing theoretical models it may be the only thing we can do. An empirical model, even if is only a mathematical function that is nothing more than a restatement in mathematical form of the actual behavior of the system, can facilitate computer processing of the observations and will be indispensable for procedures such as interpolation or extrapolation. Such models can be used, for example, to predict the response of a country's GNP to a change in taxation or to obtain measurements of temperature from the calibration curve of a resistance thermometer.

In addition, in simpler systems for which some hope of constructing a theoretical model from first principles exists, some functions, if shown to be appropriate to the observations, can give valuable guidance in model building by suggesting the type of physical process involved in the phenomenon. Even so we must be careful. The fact that we have identified a function that seems to be consistent with our set of observations at a particular level of precision does not "prove" that we have found the "right" function. Quite often functions of widely varying type can show closely similar variation, especially over a short range of the variables, and "guidance" from an inappropriately identified function can be very misleading. It can retard genuine theoretical progress for years, and the history of physics contains many examples of such failure to understand that any choice of an empirical function must be provisional.

With due attention, therefore, to the possibly limited significance of our procedures we shall describe some of the methods used. They can be quite simple in a few cases, and two of these are important because they involve functions that are of relatively common occurrence. Let us assume that we have made measurements of two variables which we can call x and y.

(a) Power Law

Consider the function

$$y = x^a$$

where a is a constant. We have

$$\log y = a \log x$$

and a graph of log y vs. log x is a straight line with slope a. Consequently if we wish to test whether a power law is a good function for our observations, we can plot them in the form log y vs. log x. If the resulting points, plotted in this way, correspond well with a straight line, we can say that a power-law function is a good fit to our observations. The value of the appropriate power, a, will be derived from the slope of the graph and will be obtained within uncertainty limits that depend on the uncertainties plotted on the points. A graph like this can be plotted on ordinary graph paper by plotting the actual values of log x and log y, or we can use logarithmic graph paper. This has rulings that are spaced in proportion to the logarithms of the numbers, so that we can plot our observations directly on the paper.

(b) Exponential Functions

For many physical phenomena an exponential function is appropriate. Consider

$$y = ae^{bx}$$

where a and b are constants. In this case

$$\log_e y = \log_e a + bx$$

and the graph of the function will be a straight line when we plot $\log_e y$ vs. x. If, therefore, there is reason to suspect that an exponential function is appropriate to a particular system, we should do a "semi-log" plot, either on ordinary graph paper by looking up the values of $\log_e y$ or on "semi-log" graph paper which has one logarithmic and one linear scale. The appropriate values of a and b will be obtainable from the intercept and slope of the line, with uncertainties determined by the plotted uncertainties of the measured values.

6–11 POLYNOMIAL REPRESENTATION

If neither a simple power nor an exponential function has been found to provide a good match to a set of observations, the probability of stumbling on a more complicated function that would be appropriate is very small. In such cases it is often useful to resort to a polynomial representation

$$y = a_0 + a_1 x + a_2 x^2 + \cdots$$

Although such a representation may not contribute much insight regarding the fundamental, theoretical basis for the operation of the system, it will at least offer some of the advantages of empirical models. If nothing more it allows machine processing of observations and provides a satisfactory basis for interpolation and extrapolation.

The coefficients of such a polynomial expansion appropriate for our particular system can be found using the least-squares principle. Recalling the remarks of Sec. 6–8, it will be appreciated that the associated difficulties escalate rapidly with the number of terms that are required in the polynomial to give satisfactory correspondence with the observations. A fuller discussion of such methods will be found in the text by Draper and Smith that is listed in the Bibliography.

A similar method is available if the scatter in the observations is not too severe and if the highest precision is not required. The techniques of the calculus of finite differences can be applied to the observations, and a difference table can be used for interpolation and extrapolation or for polynomial fitting. A complete discussion of difference-table methods will be found in the texts by Whittaker and Robinson and by Hornbeck that are listed in the Bibliography, and an elementary description will be found in Appendix 3.

6–12 OVERALL PRECISION OF THE EXPERIMENT

At the beginning of the experiment we looked forward with a guess at the uncertainties that were likely to be encountered. This was only an estimate made for the purpose of supplying guidance for the conduct of the experiment. At the end of the experiment we should look back and, by critical assessment of the results, evaluate the precision actually achieved. It does not matter very much what type of uncertainty we choose—estimated range of possible value, standard deviation, standard deviation of the mean, etc.—provided only that we clearly state the kind of uncertainty that is being quoted.

To be useful, the overall uncertainty figure must be realistic and honest, even if the outcome of the experiment is less favorable than we hoped. It should also include all identifiable sources of uncertainty. There is no point in claiming that potentials read on a 1-m slide-wire potentiometer are precise to $\frac{1}{5}\%$, simply because the scale is graduated in millimeters, if the balance point could not be identified within 2–3 mm or if errors are introduced by nonuniformities of the slide wire.

Known contributions from systematic errors should not be included at this stage, because the appropriate corrections to the measurements should already have been made. On the other hand a source of systematic error, whose presence we suspect but whose contribution cannot be evaluated accurately, should be described and appropriate allowance made in the overall range of uncertainty. The final statement will depend on the circumstances.

(a) Result Is the Mean of a Set of Readings

The best quantity to quote is the standard deviation of the mean, since this has recognizable numerical significance. Sometimes the standard deviation itself is quoted. In all cases it is essential to quote the number of readings so that the reliability of the σ estimate can be judged.

(b) Result Is the Consequence of a Single Calculation

In the undesirable event that no graphical analysis has been possible and the result is obtained algebraically from a number of measured quantities, use the methods of Chap. 3 to calculate outer limits for the uncertainty, or the standard deviation.

(c) Result Is Obtained Graphically

If the straight line has been established by a least-squares method, the uncertainties in the constants m and b are obtained directly. Note once again that these uncertainties have the advantage that they have been obtained from the actual scatter of the points, regardless of their estimated uncertainties. (This does not mean, of course, that if we intend to make a least-squares fit to a straight line, we should not bother to plot the uncertainties, or even to draw a graph at all. As was emphasized in Sec. 6–9, the graph, with the uncertainties on the points, is still needed to judge the range of matching between the model and the system before the least-squares calculation is done.) If the straight line has been drawn by eye, the lines at the limits of possibility will give the possible range of slope and intercept. This uncertainty in slope will then probably have to be combined with the uncertainties of some other quantities before the final uncertainty of the answer can be stated.

As we have mentioned earlier, it probably does not matter too much what kind of uncertainty is quoted, so long as one is quoted and the nature of the quoted value is made clear. Also, when working through lengthy uncertainty calculations, the arithmetic may be simplified by dropping insignificant contributions to the total uncertainty. There is no point in adding a 0.01% contribution to one of 5%. In the final statement of uncertainty it is not commonly valid to quote uncertainties to more than two significant figures; only work of high statistical significance would justify more.

6–13 SIGNIFICANT FIGURES

Once the overall uncertainty of the final answer has been obtained, the question of the number of significant figures to be retained in the answer can be considered. This matter has already been covered in Sec. 2–11; we repeat the discussion here simply for the sake of completeness as we discuss experiment evaluation.

There is no unique answer to the question of significant figures, but, in general, one should not keep figures after the first uncertain figure. For example, 5.4387 ± 0.2 should be quoted as 5.4 ± 0.2, because, if the 4 is uncertain, the 387 are much more so. However, if the uncertainty is known more precisely, it might be justifiable to keep one more figure. Thus, if the uncertainty above were known to be 0.15, it would be valid to quote the answer as 5.44 ± 0.15.

If a measurement is quoted with a percentage precision, the number of significant figures is automatically implied. For example, what could be meant if a measurement were quoted as $527.64182 \pm 1\%$? The 1% means that the absolute uncertainty could be calculated to be 5.2764. The precision itself, however, is quoted to only one significant figure (1%, not 1.000%), so that we are not justified in using more than one significant figure in the absolute uncertainty. We shall call the absolute uncertainty 5, and this implies that, if the 7 in the original number is uncertain by 5, the 0.64182 has no meaning. The measurement could then be quoted as 528 ± 5. If a set of readings has yielded a mean as the answer, the number of significant figures in the mean will be governed by the standard deviation of the mean, and the number of significant figures in the standard deviation will be governed, in turn, by the standard deviation of the standard deviation.

Finally, we should always be sure to quote an answer and its uncertainty in such a way that the two are consistent—i.e., neither as 16.2485 ± 0.5 nor as 4.3 ± 0.0002.

6–14 THE CONCEPT OF CORRELATION

We have, until now, been considering the interpretation of experimental results in which relatively precise observations were available and the models were relatively satisfactory. We are not always so lucky, and much of modern experimenting is less simple and clear-cut than the preceding sections might suggest. In many areas of science it is common to be concerned with subtle phenomena in which the effects we seek to measure can be wholly or partially masked by statistical fluctuation or other perturbations. In such cases, far from being able to make detailed comparisons between the system and a model, we may find it difficult to obtain clear-cut evidence that the effect we are considering even exists. This is a not uncommon situation in biological, medical, and environmental studies. We are all familiar with the discussions about the role of smoking in lung cancer, of low levels of ionizing radiation in leukemia, or of diet in cardiovascular disease. In such cases the concept of "proof" is almost always brought into the discussion in phrases like: "We have not 'proved' that smoking 'causes' lung cancer." "Can we prove that heart attacks are less likely if we eat margarine instead of butter?" In cases like these we are in a very different area of operation from our earlier kind of experimenting, and it is worthwhile spending a moment to think about what we mean by words like "proof" and "cause."

Consider two experiments. One might be a measurement of the current through a resistor as a function of potential difference across it, and the result might be as shown in Fig. 6–6(a). Have we "proved" that the current is "caused" by the potential difference? Certainly the current at the top end of the range is different from that at the low end by an amount greatly in excess of the uncertainty of measurement, and that gives us confidence that the variation actually existed. Given that it existed, was it "caused" by the change in potential difference? On that one occasion we certainly did observe that the current did increase as the potential difference increased. However, it could be that current has nothing to do with potential difference, and the increase in current was caused by some totally separate factor, like atmospheric pressure. The apparent relationship with potential difference could have been totally accidental. Philosophers for hundreds of years have been warning us that events, observed to take place simultaneously, are not necessarily causally related. In the present case, however, accumulated experience with the experiment, using repetition and careful attention to the control of other variables, will, of course, gradually convince us that potential difference and current are genuinely related, and only a philosophical purist would quarrel with the claim that the potential difference "caused" the current to flow.

The situation is quite different in less clear-cut cases. Another experiment might yield the result shown in Fig. 6–6(b). This would be likely if we were dealing with, perhaps, the number of colds experienced by the whole student body of a university as a function of the amount of ascorbic acid ingested daily. Can we say that the number of colds is dependent on ascorbic acid dose or not? We might find that, after performing a well-designed experiment using an experimental group and a control group [as described in Sec. 5–6(b)], a control group of 100 subjects who unknowingly swallowed sugar pills instead of ascorbic acid every morning had 125 colds while the experimental group

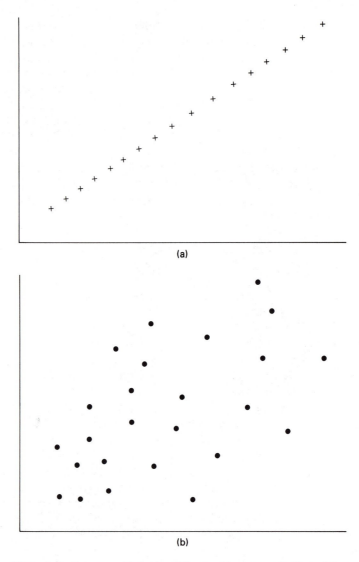

Figure 6–6   Extremes of behavior with respect to cause-and-effect relationships.

who were taking ascorbic acid had a total of 106 colds. The questions we must ask are: "Is this difference 'significant'?" "What do we mean by 'significance'?" "If the difference is significant, can we ascribe it to the ascorbic acid?"

Even painstaking attention to the details of experimenting, control over samples, elimination of extraneous variables, repetition of the experiment, etc. may not clear up the situation very much. Biological systems are sufficiently complex that we can rarely attain the degree of control over variables that characterized the electrical experiment. It therefore becomes inappropriate to seek the kind of "proof" that is available in other systems. We cannot say that we have "proved" that smoking causes lung cancer or that ascorbic acid reduces the incidence of colds in the same way that we can "prove" that a potential difference "causes" a current to flow. We have to be content with another class of statement, which, although less exact, can still be adequately significant and completely convincing.

This type of statement can be illustrated by reference to a diagram such as Fig. 6–7. These measurements were made to test the proposition: the number of counts obtained from a weak radioactive source depends on the length of time of counting. Here statistical fluctuation is almost as big as the effect we

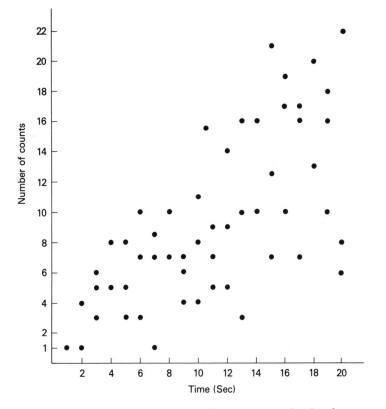

Figure 6–7   The dependence of number of counts on counting time for a weak radioactive source.

seek to observe, but we can still see that there is an upward trend in the observations. In such a case we say that there exists a "correlation" between one variable and the other. This means that we can observe a tendency for one variable to follow the other, although fluctuations arising from other factors prevent the observation of a unique, one-to-one correspondence. The mathematical study of such correlation is called "regression analysis." Regression analysis provides a numerical measure of the degree of correlation between the two variables, and we can evaluate, for a particular set of observations, a "correlation coefficient." The value of this quantity will depend on the degree of scatter of the observations, varying from 1, if all the observations should fall exactly on a line, to 0, if there is no relationship between the variables at all.

We encounter the concept of correlation in two significant cases: (a) if, of two measured variables, one can be regarded as the cause of the other but its effect is partially masked by random fluctuation, and (b) if two variables can be regarded as simultaneous consequences of a common cause whose effect, as before, is partially obscured by random fluctuation. In either case we may be able to say that we can observe a certain degree of correlation between one variable and the other.

The mathematical properties of correlations are described in the standard texts on statistics. We shall confine ourselves here to quoting the equation by which one calculates correlation coefficients. For a pair of measured variables x and y the equation is

$$r = \frac{\Sigma\, xy}{\sqrt{\Sigma\, x^2\, \Sigma\, y^2}}$$

$$\frac{\Sigma (x-\bar{x})(y-\bar{y})}{\sqrt{\Sigma (x-\bar{x})^2\, \Sigma (y-\bar{y})^2}}$$

Once calculated, such a correlation coefficient will supply a very important figure, namely a measure of the extent to which the observed variation in one quantity can be ascribed to variation in another. In case (a) above it will be the extent to which the variation in the output variable can be ascribed to variation in the input variable; in case (b) it will be the extent to which the variation in both variables can be ascribed to variation in whatever is the common source of influence.

Even when we do observe correlations, we must still be very careful about inferring causal connection between the various variables. If we observe that one variable seems to correlate well with another, we have not "proved" that one variable "causes" the other in the same sense as we used these words in the example about electric current. The literature has many examples of false and misleading correlations. One conference speaker illustrated this point with a tongue-in-cheek claim to have discovered the cause of cancer. He showed a graph of a quantity which correlated beautifully with the increase in some type of cancer and only later revealed that the other variable was the consumption of fuel oil in the British Navy. In another case, intended apparently to be taken seriously, a 1920s newspaper report described the discovery of the cause of polio, since the incidence correlated so well with the number of motor cars on the roads.

However, such amusements do not discredit the study of correlations or the search for causal relations; they merely serve as another reminder of the need for caution and clear thinking. When treated with great care, and especially when the correlation can be observed repeatedly, correlation studies can, and do, supply convincing evidence of causal connection. Because of the immense importance which many of such issues have in public affairs it is important for all of us to have a clear understanding of the nature of correlation and the methods available for significance testing. Further discussion is beyond the scope of this volume, but pursuit of the topic in the texts on statistics (listed in the Bibliography) is earnestly recommended.

PROBLEMS

1. An experiment was done to measure the impedance of a series R-L circuit. The impedance Z is given as a function of the resistance R, the frequency of the source f and the inductance L by

$$Z^2 = R^2 + 4\pi^2 f^2 L^2$$

The experiment was done by measuring Z as a function of f with the intention of plotting Z^2 vertically and f^2 horizontally to obtain L from the slope and R from the intercept. The observations obtained are given in the table.

f, Hz	Z, Ω	f^2	$f(AUf)$	AUf^2 $= 2f(AUf)$	Z^2	$Z(AUZ)$	AUZ^2 $= 2Z(AUZ)$
123 ± 4	7.4 ± 0.2						
158	8.4						
194	9.1						
200	9.6						
229	10.3						
245	10.5						
269	11.4						
292	11.9						
296	12.2						

The uncertainties given in the first line refer to all the readings in each column.

(a) Plot these readings in the appropriate fashion and mark the uncertainties on the points. Suggested table headings to expedite the calculations are given above.

(b) Check to see if the observations can be interpreted in terms of a straight line for any part of the range or all of it.

(c) Obtain the slope of the best line.

(d) Calculate the best value for L.

(e) Obtain the slopes of the lines at the outer limits of possibility and so state the range of uncertainty for the slope.

(f) Calculate the absolute uncertainty in the measurement of L.

(g) Obtain the best value of R from the intercept.

(h) Obtain the uncertainty for the R value.

(i) State the complete result for the experiment with the appropriate number of significant figures in each quantity.

2. Ten different observers report on the intensity of a lamp measured with a comparison photometer. Their results (in arbitrary units) are as follows:

Observer	Mean	Standard deviation of mean
1	17.3	2.1
2	18.4	1.9
3	17.1	2.5
4	16.6	2.8
5	19.1	3.2
6	17.4	1.2
7	18.5	1.8
8	14.3	4.5
9	16.8	2.3
10	17.4	1.6

What is the result of the experiment and what is its standard deviation?

3. An experiment has been carried out to investigate the temperature dependence of the resistance of a copper wire. A common model is represented by the equation

$$R = R_0(1 + \alpha T)$$

where R is the resistance at temperature $T°C$, R is the resistance at 0°C, and α is the temperature coefficient of resistance. Observations of R and T were obtained as given in the table at the top of p. 135.

(a) Using the method of least squares (suggested headings under which the calculations can be carried out are given in the first part of the table), obtain the best value for the slope and for the intercept.

(b) Hence obtain the best value for α.

(c) Evaluate the standard deviation for the slope and for the intercept (suggested

Calculation of best values of R_0 and α				Calculation of standard deviations using best values of R_0 and α			
x	y	xy	x^2				
$T°C$	$R\ \Omega$	TR	T^2	$R_0\alpha T$	Calculated ideal value of R ($=R_0$ $+ R_0\alpha T$)	δR (obs. R $-$ ideal R)	$(\delta R)^2$
10	12.3						
20	12.9						
30	13.6						
40	13.8						
50	14.5						
60	15.1						
70	15.2						
80	15.9						
$\Sigma\ x$ and $(\Sigma\ x)^2$	$\Sigma\ y$	$\Sigma\ (xy)$	$\Sigma\ (x^2)$				$\Sigma\ (\delta y)^2$ and so s_y

headings for this part of the calculation are given in the second part of the table).

(d) Hence evaluate the standard deviation of α.

(e) State the final result of the experiment with the appropriate number of significant figures.

4. It is desired to fit a set of observations to the function $y = a + bx^2$ using least squares. Use the same procedures that are used in Appendix section A2–2 for calculating the constants of a linear function to obtain equations for a and b in the parabolic function. Hence calculate the values of a and b appropriate to the following set of observations:

x	y
0.5	1.5
1.0	6.3
1.5	12.4
2.0	12.6
2.5	18.0
3.0	32.8
3.5	40.2
4.0	47.4

Assume that uncertainty is confined to the y variable.

5. The following measurements were made in the investigation of phenomena for which no existing model was available. In each case identify a suitable function and evaluate its constants.

v	i
0.1	0.61
0.2	0.75
0.3	0.91
0.4	1.11
0.5	1.36
0.6	1.66
0.7	2.03
0.8	2.48
0.9	3.03

(a)

x	y
2	3.2
4	16.7
6	44.2
8	88.2
10	150.7
12	233.5
14	337.9
16	464.5
18	618.0

(b)

T	f
100	0.161
150	0.546
200	0.995
250	1.438
300	1.829
350	2.191
400	2.500
450	2.755
500	2.981

(c)

(This last one is a little less obvious)

6. The following results come from a study on the relationship between secondary-school matriculation averages and the students' overall average at the end of first-year university. In each case the first number of the pair is the secondary-school average and the second is the university average.

    78,65; 80,60; 85,64; 77,59; 76,63; 83,59; 85,73; 74,58; 86,65;

    80,56; 82,67; 81,66; 89,78; 88,68; 88,60; 93,84; 80,58; 77,61;

    87,71; 80,66; 85,66; 87,76; 81,64; 77,65; 96,87; 76,59; 81,57;

    84,73; 87,63; 74,58; 91,78; 92,77; 85,72; 86,61; 84,68; 82,66;

    81,72; 91,74; 86,66; 90,68; 88,60.

    (a) Draw a "scatter" diagram of university average plotted against school average.
    (b) Evaluate the correlation coefficient.

7. Evaluate the correlation coefficient for the values of $\sqrt{x}$ and t in Table 4–3.

7

Writing Scientific Reports

7–1 GOOD WRITING DOES MATTER

It is almost impossible to overestimate the importance of good scientific writing. The best experimenting in the world can be of little or no value if it is not communicated to other people—and communicated well by clear and attractive writing. Although communication may sometimes be verbal, in the overwhelming majority of cases people will learn about our work from the printed page. Our obligation to become as literate as we can is therefore not trivial, and it should be regarded as an essential and integral part of our experimenting activities. Our writing must be sufficiently good that it will attract and retain the interest and attention of our readers. This chapter will contain some hints on how this may be achieved.

It is almost impossible to tell a person how to write well. It would be very convenient if we could lay out a short list of instructions and guarantee thereby fluent, lucid, and literate prose, but the list is not available. Each of us has different ways of expressing our thoughts, and each must allow his or her writing style to develop in its own way. This needs extensive practice, and we should regard report writing in the introductory physics laboratory as an excellent opportunity to obtain it. We may end up acquiring different writing styles, but, provided the message is clear, the diversity can be enriching rather than damaging.

Let us now turn to some practical considerations concerning the actual writing of reports. Although we have noted that we cannot produce a list of explicit instructions for good writing, there does exist one principle which, if we follow it in our own ways, will make it much more likely that we shall produce good, readable prose. Whether we are preparing a report for internal circulation in a private organization or a paper for publication in the open literature, there is one person whose interests must claim our first attention—the person who is actually engaged in reading our report. That person is, as far as our report is concerned, the most important person in the world, and we are well advised to concentrate our attention on him or her. Our readers are very likely people we do not know, perhaps in some far-distant part of the world and very likely knowing nothing about us or our work except the report that they hold in their hands. We probably have only one chance to influence them—as they read our report—and the report must do it alone. We cannot stand at our reader's shoulder, adding explanation and clarification as he encounters difficulty in understanding what we have written. Not only must the report stand on its own, but the outcome of the reading can be highly significant. The public recognition of our work, the opportunity of others to benefit from it, our own reputation, perhaps even our chances of employment or promotion may depend on these few minutes as our reader works his way through our report. Do we have to be persuaded further that we should take writing seriously?

With respect to language it has been common in the past to recommend a detached, depersonalized mode of expression that was characterized by the use of the passive voice and impersonal constructions. There seems to be little point in perpetuating such stilted language, and we can simply tell our reader what we did in our experiment, e.g., "We measured the time of fall using an electronic timer that was accurate to 1 millisecond." Since there is no single "right" way to write an experimental report, we should feel free to use such language and modes of expression as allow us to express our thoughts in the most clear, attractive and persuasive way possible. For invaluable advice regarding writing style consult the little book by Strunk and White that is listed in the Bibliography.

We shall now consider the various sections of the report in turn, all as seen through the eyes of our all-important reader.

7–2 TITLE

The title is probably the first part of the report to draw the attention of our readers. Since they are almost always busy people with many items competing for their attention, our report will capture their interest only if the title is informative, appropriate, and attractive. It should not be too long, and yet should specify quite explicitly the topic of the work. For example, if the purpose of our experiment is to measure the specific heat of a fluid using continuous-flow calorimetry, we could use this directly as a title: "Measurement of the specific heat of water using continuous-flow calorimetry." It may be helpful to note that three questions are answered in this title: (a) Is the work experimental or theoretical? That is, are we reporting a measurement or a calculation? (b) What is the topic of the work? (c) What general method did we use? Attention to these three items will almost invariably result in a good choice of title. One useful hint is to avoid the word "the" as the first word in a title. Titles are sometimes included in lists in alphabetical order, and readers may find it harder to identify our work if there are too many entries starting with "The."

7–3 FORMAT

The sections that follow offer some detailed analysis of the various parts of a report. The various subsections to be described should not be used as headings in actual reports. While practice will obviously vary with circumstances, reporting on most normal work in the introductory physics laboratory will need only the minimum of division. The sections of a report that are essential are

> INTRODUCTION
> PROCEDURE
> RESULTS
> DISCUSSION

and these should be used as a basic starting point. The section headings should be neat, clear, and written in block capitals. Subsections within each of these main sections should be used only when the length or complexity of the report makes them indispensable for clarity. Other main sections may, of course, be introduced in accordance with the requirements of particular experiments. Suggested possibilities are:

> THEORY
> SAMPLE PREPARATION
> UNCERTAINTY CALCULATION
> etc.

To make the report as inviting to read and as easy to understand as possible, it should contain a clear, logical thread of argument, and we should not allow anything to disrupt that development of thought. If we feel that we must include some particular piece of description that is so lengthy and detailed that it would interrupt the smooth development of the main argument, we should consider making it an appendix to the report. In that way all the detail is available to any reader who wants it, but the main continuity of thought is not broken.

Let us turn now to the details of each section of the report.

7–4 INTRODUCTION

(a) Topic Statement

If our title has been successful, we can assume that we have now attracted our reader's attention and he has picked up our report. However, he is almost certainly starting from zero, or close to it, as far as our particular experiment is concerned, and, as he starts to read, our first task is to orient his thinking toward our particular area of study. We are not going to succeed in this by diving immediately into unorganized detail about the experiment. Think instead of the most general statement that can be made about the experiment and state it directly—e.g., "It is possible to measure the gravitational acceleration using the oscillation of a simple pendulum." In this way our reader is taken from his initial state of ignorance to direct awareness of the specific topic of our work.

(b) Review of Existing Information

The natural reaction of our reader at this point will be to hope for some reminder of the basic information relating to this particular area. We can meet this need by giving him a brief summary of the existing state of knowledge relevant to our experiment. This may include, as necessary, some aspects of the history of the subject and/or a summary of earlier experimental work. In addition, however, two items are not discretionary and must be included in every report on an experiment. One is a clear statement of the system and the experimental circumstances with which we are dealing, and the second is a description of the model or models that we are using.

It is generally best to give this summary quite briefly for fear of obscuring the main line of argument, but it should be sufficiently detailed that reference is made to all aspects of the situation that will be necessary for understanding the rest of the report. In the interests of brevity and clarity, however, the derivation of standard theoretical results associated with the model should not be included. (The way in which these standard results are manipulated to refer to our particular system is another matter, however, because that is specific to *our* experiment, and this will be the topic of later discussion.) The behavior of the model, as represented by important equations, should be quoted, and it is important at this stage to include mention of any assumptions contained in the model that may limit the validity of the equations—e.g., "It can be proved that, in the limit of vanishingly small amplitude of oscillation, the period of oscillation of a simple pendulum, considered to be a point mass at the end of a massless, inextensible string, is given by . . . ." To compensate for the omission of standard derivations we may feel it desirable to include in our references a source in which the complete derivation will be found.

(c) Application of Information to Specific Experiment

On the basis of the material of the preceding section our reader will be equipped to understand all that follows in the report, and his natural reaction at this point will be to wonder: how does all this refer to this particular experiment? We should, therefore, supply a paragraph or two to show how the basic information, such as an equation representing the behavior of the model, can be converted to provide a foundation for our particular experiment. Commonly this will involve some procedure such as putting the basic equation into straight-line form (or some suitable equivalent) and identifying the ways in which the model can be tested against the system. We can also point out at this stage the information that will become available from the parameters of the graph (such as slope and intercept in the case of straight-line plotting). Our reader will thereby become fully aware of the way in which our final answer will be obtained.

(d) Summary of Experimental Intention

It is very satisfying to the reader if we conclude our introduction with a summary of our specific intention in the experiment. For example "Thus, by measuring the variation of index of refraction with wavelength, we shall be able to test Cauchy's model using a graph of n vs. $1/\lambda^2$. The values of Cauchy's coefficients A and B that are appropriate to our glass specimen will then be obtained, respectively, from the intercept and slope of the graph." Such a statement is satisfying to the reader because, particularly in a long and complicated experiment that required a lengthy introduction, it offers him a review, in summary form, of the whole course of the experiment, thereby equipping him to proceed to a description of its actual conduct.

(e) Statement of Experimental Purpose

You may have noticed that no mention has yet been made of the traditional statement of purpose for the experiment. It has been omitted so far because, although it certainly should appear somewhere in the introduction, there is no universally suitable location. If the topic of the experiment is familiar, the statement of purpose could form an acceptable topic statement, right at the beginning of the introduction—e.g., "It is the purpose of this experiment to measure the acceleration of gravity by timing the fall of a freely falling object." Under suitable circumstances such a statement of purpose can make an excellent topic statement. On the other hand the basic purpose of an experiment might involve matters so complicated and unfamiliar that a statement of it would be completely incomprehensible unless it followed a substantial amount of preparatory material. It is easy to imagine a complicated theoretical description that could profitably conclude with the phrase: ". . . and it is the purpose of this experiment to evaluate coefficient k in equation 10." It does not matter a great deal, therefore, where our statement of purpose comes, so long as it is included and comes at a point in the introduction where it fits well and makes good sense to the reader.

　　　The introduction has performed a number of services for our reader. Right at the beginning the topic statement has directed his attention to our particular area of work. He has then been reminded of the existing state of knowledge in that area. Next he has been shown how that applies to our particular experiment. Finally, he has been given a concluding summary of our specific experimental intention. He is now ready to hear how we actually did the experiment.

7–5 PROCEDURE

You may have noticed that the report's introductory section takes the form of a descriptive sequence proceeding from general to specific. We start with a topic statement that is the most general remark about the experiment we can make, and we end with a completely specific statement of intention. Such a sequence is designed to suit our reader's requirements in the introductory section, and a similar sequence will be equally suitable for the procedure section. If we were to start our description of procedure by launching immediately into a mass of unorganized detail, we would succeed only in irritating our reader. How can he appreciate the great care we took in some experimental detail if he is not aware even of our measured variables? We should be as considerate of our reader's mental efforts as we write our procedure section as we were in the introduction, and a second sequence from general to specific is clearly called for.

(a) Outline of Procedure

To set the scene for our subsequent description of the details of procedure and measurement we should first offer our reader a review of the whole course of the experiment. If the experiment really consisted of the measurement of the variation of electrical resistance of a copper wire with temperature over the range 20°C to 100°C, we should say just that. It will provide the reader with a framework into which he can fit all our subsequent description of detail. If we start our description of procedure by saying that we connected terminal A to terminal B, switched on power supply C, read voltmeter D, . . . etc., we shall have lost his attention in two lines.

(b) Specific Measurement Details

Now that our reader knows the general course of the experiment, he is ready to be told the specific methods by which we measured each of the required quantities, carried out sample preparation, etc. This can be done quite simply by stating each in turn untill we have completed the list. We must make sure that no significant method of measurement is omitted; in such a thing as a timing measurement it is almost certainly important that we used an electronic timer with millisecond accuracy rather than a stopwatch that could be read to $\frac{1}{5}$ sec, and our reader will want to know that we did so. If a quantity in the experiment could be measured using some standard and familiar technique, it may be sufficient to mention it by name — e.g., "The resistances were measured using a Wheatstone bridge accurate to 0.01%." If we feel it is unusually significant, we can discuss at this stage the accuracy of any particular measuring process, while remembering that the overall precision of the experiment is a different topic that will appear in a subsequent section of the report.

(c) Precautions

Once our reader has learned the methods by which we made each of our measurements, he may recall the difficulties or opportunities for error that are inherent in particular procedures. He would, therefore, appreciate reassurance that we, too, had thought of these possibilities and had been sufficiently careful in taking the necessary precautions. As we offer it, however, we need not go to extremes; care should obviously be taken with all measurements, and there is no point in troubling our reader with superfluous claims to virtue in describing routine and obvious precautions. There are times, however, when special care to avoid some particular source of error is a genuinely important part of the experiment, and it is perfectly reasonable to draw attention to this before we close our procedure section.

(d) Apparatus Diagrams

Good diagrams of experimental apparatus are a virtually essential part of any good report. While a published paper will require drawings of professional quality, and such resources are not available in introductory work, we should

early acquire the habit of taking care with apparatus drawings. Even if sophisticated drawing aids are not available, it is not too much to expect the use of a
ruler. Neatness and clarity will be much appreciated by our reader, and good,
legible labelling will assist him enormously in understanding our experiment.
Good diagrams can help us, too, as we write our report. Reference to a good,
clear, well-labelled diagram may save us paragraphs of written description and
provide detail that would be intolerably tedious to read if it were included in
the text.

Reference to diagrams can, of course, be inserted at any appropriate
point in the text, but reference to a general diagram of our apparatus as a whole
can make a neat and convenient beginning for a procedure section—e.g.,
"Using the apparatus shown in Fig. 1, we measured the variation of time of
fall of a ball bearing with height over the range 20 cm to 150 cm." An example
of an acceptable apparatus diagram will be found in Fig. 7–1.

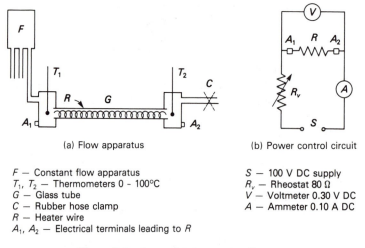

(a) Flow apparatus   (b) Power control circuit

F — Constant flow apparatus S — 100 V DC supply
T_1, T_2 — Thermometers 0 - 100°C R_v — Rheostat 80 Ω
G — Glass tube V — Voltmeter 0.30 V DC
C — Rubber hose clamp A — Ammeter 0.10 A DC
R — Heater wire
A_1, A_2 — Electrical terminals leading to R

Figure 7–1   A completed apparatus diagram.

7–6 RESULTS

(a) Measured Values

At this point the reader has been given all the information he needs for understanding the experiment and he is ready to receive the results directly. Since
any good experiment will almost inevitably involve the variation of some
quantity with another, the results are almost certainly best presented in a table.
Here, as always, neatness and clarity are of paramount importance. Lines for

the table should be drawn with a ruler, and ample space should be provided for the headings and for the columns of figures. The headings should be explicit and should include, if possible, the name of the variable, its symbol, and the units of measurement. Attached to each numerical entry should be its uncertainty, unless some separate discussion of uncertainties makes the precision of the measurements absolutely clear. The table or tables should be clearly identified with a table number and a clear heading or caption. At this point it may be appropriate to refer to any graphs we have drawn of the basic variables. A simple statement like "A graph of the time of fall vs. height will be found in Fig. 2" will suffice. Any tables of values that are so extensive and detailed that their inclusion in the main text of the report would disrupt the train of thought should be relegated to an appendix.

Following the main table or tables we should list the measured values of all the other quantities in the experiment. As always, each should have its uncertainty attached, and the units of measurement should be clearly stated.

(b) Description of Measurement Uncertainties

We should state explicitly the kind of uncertainties we are quoting. They will very likely be either estimated outer limits or statistical quantities such as a standard deviation or a standard deviation of the mean. In the case of statistical quantities we must not omit mention of the number of readings in the sample from which the results were derived. If any quantities in our list of measured values were obtained by computation from some basic measurement or measurements, we should state clearly the type of calculation used to obtain the final uncertainty in the computed quantity. In such cases it is not necessary to give much, if any, of the arithmetic details, provided that our reader can see clearly the kind of calculation we performed.

(c) Computation of Final Answer

If our experiment has been well designed, we shall almost certainly be obtaining our final answer by some graphical procedure. It is now time to tell our reader exactly what that procedure is. In simple cases we may obtain our answer from the basic graph of one measured variable against the other, but, even then, we must tell our reader explicitly what we have done—e.g., "The value of the resistance was obtained from the slope of the graph (shown in Fig. 3) of V vs. I between 0.5 A and 1.5 A." If our answer was not the slope itself but was obtained from calculation with other measured quantities, we should, once again, state explicitly what we have done—e.g., "Our value for coefficient of viscosity was obtained from the slope of the graph of Q vs. P in combination with the measured values of a and ℓ using Eq. (3)."

At this stage our reader will wonder what kind of calculation we performed to obtain the uncertainty in the final answer. We should simply say what we have done. If we assessed visually the possible range of slopes, we should say just that. We can add, if necessary, that the basic uncertainty in slope was combined with other uncertainties, and state explicitly the method of calculation. If we obtained our slope by a least-squares calculation and incorporated any other standard deviations to obtain a final value for the uncertainty of the answer, we should, again, state simply what we have done.

Throughout our results section we should not trouble our busy reader with unnecessarily detailed calculations. He will trust us to do simple arithmetic, but he will want to know what kind of calculation we did. If we feel compelled, for some particular reason, to offer an unusual amount of detail regarding such calculations, we can always put it in an appendix where it will be available if wanted but will not obscure the clarity of our main report.

7–7 GRAPHS

The graph, or graphs, in the report differ from the graphs we used in doing our experiment. Those graphs were working documents that were designed as computational aids. For a precise experiment the graphs were possibly quite large, and they should have been finely drawn to permit precise extraction of information. On the other hand, it is extremely unlikely that our reader will want to do any numerical work of his own using the graphs in our report. These graphs serve mostly as illustrations. They allow the reader to see the behavior of our system and thereby enable him to judge for himself the validity of our assertions about the results.

The graphs in our report should, of course, be clear, neat, and uncluttered so that the reader does not have to work too hard to get their message. The points on the graphs should have their uncertainties clearly marked on them (by a box or a cross) and the axes should be clearly labelled. Both the type of uncertainty and any symbols used in labelling the axes should be explicitly identified in some obvious way in or beside the graph; we do not want to cost our reader the irritation of hunting through the text to find out how to interpret the graph. Do not, however, fill up empty spaces on the graph with arithmetic calculations of slopes, etc. Each graph should obviously have a clear title, or, as is common in printed publications, a more extended caption. In addition to supplying identification, an extended caption has the added advantage of supplying a good location for the important details mentioned above. A sample of acceptable layout for such an illustrative graph is given in Fig. 7–2.

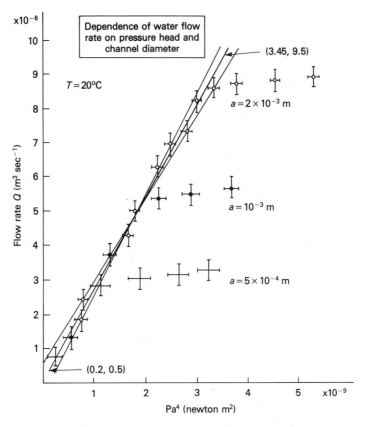

Figure 7–2 A completed graph.

7–8 DISCUSSION

(a) Comparison Between Model and System

The discussion is an integral part of the report and not an afterthought. It is so because we have yet to describe the aspect of the experiment that we have, right from the beginning, regarded as the basic issue in experimenting—the relationship between the system and the model. The outcome of that comparison is vital for the experiment, and our reader will be eager to hear what we have to say about it.

We have already listed in Sec. 6–11 the various categories of experimental outcome. Remembering that, in evaluating our results, we had to disengage ourselves from our hopes and aspirations for the experiment and accept objectively the actual outcome, so now, at the reporting stage, we must make a candid and unbiased statement of that outcome. We should make it a plain, simple statement of the actual situation—e.g., "The behavior of the model is represented by Eq. (1), in which the variation of Q with P is a straight line passing through the origin. In our experiment the results did show a linear variation over part of the range, but, instead of passing through the origin, the line that best fits the observations has a finite intercept on the Q axis. In addition, at the high end of the P scale, the observations show systematic deviation from linearity by an amount clearly in excess of the uncertainty of measurement."

At this stage it is sufficient to make this plain, factual statement. Since such comparison was our fundamental experimental objective, it is necessary for its outcome to be clearly, factually, and prominently stated. In our report we shall be proceeding quite soon to matters of interpretation and opinion, and it is important that we start our discussion section with a plain statement of the actual, indisputable outcome of the experiment.

That statement will raise some questions in the mind of our reader and we must turn our attention to them now.

(b) Consequences of Discrepancies Between Model and System

One of these questions will concern the possibility of error in our final answer that would be caused by failure of correspondence between our system and the model we used. Some of these possibilities have already been mentioned in Sec. 4–5, and our reader will appreciate reassurance that we have protected our final answer from that kind of error. We should point out, for example, that an unexpected intercept will not contribute to error in a quantity that has been obtained from the slope alone, or that a systematic departure from linearity over part of a graph did not invalidate an answer that was obtained from the linear segment alone. Much of the skill in experimenting lies in the protection of our final answers from such sources of error, and we can be quite explicit about our claims to have done so; our reader will appreciate the reassurance.

(c) Speculation Concerning Discrepancies Between System and Model

In describing the report's earlier sections we have stressed objective and factual reporting of the actual situation. Matters of opinion or conjecture should not have played a significant role in those parts of the report, and we have probably limited ourselves to such statements as would have been made by most impartial observers. Now, however, comes a stage at which we not only can, but should, introduce our own ideas. Our reader has, in turn, been informed about the actual degree of correspondence between our system and the model, he has been reassured that our final answer has not been contaminated (as far as we are able to tell) by any failure of correspondence between the system and the model, and we could, quite justifiably since we have met our basic obligations as experimenters, leave the report there. However, the interest of our reader will doubtless have been aroused by our description of any discrepancies between the model and the system. We presumably started with a model that was chosen to suit the system as accurately as possible. If any failure of correspondence between the system and the model had been anticipated, such breakdown would have been incorporated into our experiment design. If we want to measure a coefficient of viscosity using theories based on streamline flow, for example, we do not design our experiment to run under turbulent conditions (unless, for some separate purpose, we wish to detect the onset of turbulence). An observed failure of correspondence, therefore, is bound to attract attention, and our reader will want to know what we think about it. We are more familiar with our experiment than anyone else, and we should be in a better position than others to guess at the origin of discrepancies.

Sometimes a discrepancy will have (at least superficially) an origin that is easy to identify. If we have been measuring, for example, the flow rate of fluid through a pipe, a departure of the flow-rate measurements below linear behavior at high values of pressure difference may be ascribed quite confidently to the onset of turbulence. If our experimental objective included the detection of the onset of turbulence, such a statement could end the matter. At other times, however, a little more comment is called for. If, in the preceding example, our intention had been simply to measure the coefficient of viscosity from the linear part of the Q, P variation, our reader might wonder why we had not been more successful in avoiding a region in which the streamline theory was clearly invalid. Perhaps we had been surprised by an unexpectedly early onset of turbulence, and, if so, we should be candid enough to admit it and perhaps speculate on the origin of that discrepancy. If, finally, we are dealing with a situation that is genuinely puzzling, we may not be able to offer much in the way of speculation, but it is always worth trying. As we said before, we have a better chance of speculating fruitfully than most others, and our ideas will almost certainly be of interest and possible value to other workers.

Sometimes, however, despite our best efforts we shall fail and be unable to offer any constructive ideas. At this point we should be completely honest. If we are dealing with a well-tested system and a well-known, reliable model and we have honestly tried and failed to resolve a failure of correspondence between them, our situation cannot but be of interest to other workers. We should tell them about it, and perhaps we shall all learn something from the resulting discussion.

As we attempt to be creative regarding our experimental discrepancies, we should remember that we are doing something important. All models and theories go through processes of refinement, and these processes are based on the various stages of observed failure of the models. We should try to be responsible, therefore, as we speculate. Instead of having a fling at every wild idea we can imagine, we should try to make our suggestions have some logical connection with the evidence of the discrepancy. For example, if we have been doing an experiment on the oscillation of a load suspended from a spring, we could write: "Since the unexpected intercept in the plot of T^2 vs. m gives a finite value of T for $m = 0$, we have a clear indication of the presence of an extra mass that was not included in the measured values of load." Whether we can guess at the identity of this extra mass is less important; we have, at least, offered a logically acceptable inference from the observed nature of the discrepancy, and further research and experimenting in this area will have been facilitated.

Appendix 1

Mathematical Properties of the Gaussian or Normal Distribution

A1–1 THE EQUATION OF THE GAUSSIAN DISTRIBUTION CURVE

Consider that a quantity, whose true value is X, is subject to random uncertainty. Consider that the uncertainty arises from a number, $2n$, of errors each of magnitude E and equally likely to be positive or negative. The measured value x can then range all the way between $X - 2nE$, if all the errors should happen to have the same sign in the negative direction, to $X + 2nE$ if the same thing happened positively. Intermediate values will arise, obviously, from various combinations of positive and negative contributions. We wish to determine the form of the resulting distribution curve for a very large number of such measurements. This form will be determined by the probability of encountering a particular error R within the total interval $\pm 2nE$. This probability is governed by the number of ways in which a particular error can be generated.

For example, an error of the total value $2nE$ can be generated in only one way—all the elementary errors must have the same sign simultaneously. An error of magnitude $(2n - 2)E$, on the other hand, can occur in many ways. If any one of the elementary errors had been negative, the total error would have added up to $(2n - 2)E$, and this situation can arise in $2n$ different ways. An error of $(2n - 2)E$ is, therefore, $2n$ times as likely as one of $2nE$. A situation in which two of the elementary errors have negative signs can, correspondingly, be generated in many more ways than for one, and so on.

This argument can be generalized, therefore, by using the number of ways in which a specific error R can be generated as a measure of the probability of the occurrence of the error R and, consequently, as a measure of the frequency of its occurrence in a universe of observations.

Consider a total error R of magnitude $2rE$ (where $r < n$). This must be the result of some combination of errors of which $(n + r)$ are positive and $(n - r)$ are negative. The number of ways in which this can happen can be calculated as follows. The number of ways of selecting any particular arrangement of $2n$ things is $(2n)!$. However, not all of these arrangements are different for our purpose, since we do not care if there is any internal rearrangement between the errors in, say, the positive group. We must, therefore, divide the total number of arrangements by the number of these insignificant rearrangements, i.e., by $(n + r)!$. Similarly we must divide by the number of internal rearrangements possible in the negative group, i.e., by $(n - r)!$. The total number of significant combinations is, therefore,

$$\frac{(2r)!}{(n + r)!(n - r)!}$$

This quantity is not yet strictly a probability, although it is a measure of the likelihood of finding such a total error. The probability itself will be obtained by multiplying the above number by the probability of this combination of $(n + r)$ positive and $(n - r)$ negative choices. Since the probability of each choice is $\frac{1}{2}$, the required multiplier is

$$\tfrac{1}{2}(n+r) \; \tfrac{1}{2}(n-r)$$

The final result for the probability of the error R is then

$$\frac{(2n)!}{(n + r)!(n - r)!} \left(\frac{1}{2}\right)^{(n+r)} \left(\frac{1}{2}\right)^{(n-r)} \tag{A1--1}$$

Our problem is now to evaluate this as a function of the variable r. This is done subject to the condition that n is very large, in fact tending to infinity.

The evaluation requires two auxiliary results.
1. The first auxiliary result:

$$n! \approx \sqrt{2\pi n} \; e^{-n}n^{n}$$

This is known as *Stirling's theorem*. Although its full derivation is beyond our scope, its plausibility can be indicated as follows:

$$\int_{1}^{n} \log x \; dx = [x \log x - x]_{1}^{n}$$
$$= n \log n - n + 1$$

The graph of log x vs. x is shown in Fig. A1–1. Clearly the integral above can be approximated by the sum

$$\log 1 + \log 2 + \log 3 + \cdots + \log n$$

which is $\log (1 \times 2 \times 3 \times \cdots \times n)$ or $\log n!$. We can, therefore, write approximately, if n is large,

$$\log n! = n \log n - n$$

i.e.,

$$n! = e^{-n}n^n$$

This is an approximation to the formula given above. A full derivation will be found in the text by Margenau and Murphy listed in the Bibliography.

2. The second auxiliary result:

$$\lim_{n \to \infty}\left(1 + \frac{1}{n}\right)^n = e$$

The expansion for $[1 + (1/n)]^n$ is

$$1 + \frac{n}{1!}\frac{1}{n} + \frac{n(n-1)}{2!}\left(\frac{1}{n}\right)^2 + \frac{n(n-1)(n-2)}{3!}\left(\frac{1}{n}\right)^3 + \cdots$$

As n becomes larger, all the terms in n clearly tend to unity, so that the series tends to

$$1 + \frac{1}{1!} + \frac{1}{2!} + \frac{1}{3!} + \cdots = e$$

as required.

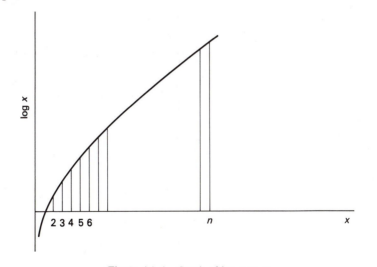

Figure A1–1   Graph of log x vs. x.

We are now in a position to evaluate the expression (A1–1). Apply Stirling's theorem to the terms $(2n)!$, $(n + r)!$, and $(n - r)!$:

$$(2n)! = (2n)^{2n} e^{-2n} \sqrt{2\pi \cdot 2n} = 2^{2n} n^{2n+(1/2)} e^{-2n} \sqrt{4\pi}$$

$$(n + r)! = (n + r)^{n+r} e^{-(n+r)} \sqrt{2\pi(n + r)}$$

$$= n^{n+r+(1/2)} \left(1 + \frac{r}{n}\right)^{n+r+(1/2)} e^{-n-r} \sqrt{2\pi}$$

$$(n - r)! = n^{n-r+(1/2)} \left(1 - \frac{r}{n}\right)^{n-r+(1/2)} e^{-n+r} \sqrt{2\pi}$$

Therefore,

$$(n + r)!(n - r)! = n^{2n+1} \left(1 - \frac{r^2}{n^2}\right)^{n+(1/2)} \left(1 + \frac{r}{n}\right)^{r} \left(1 - \frac{r}{n}\right)^{-r} e^{-2n} \cdot 2\pi$$

The variable part of (A1–1) can now be written

$$\left(1 - \frac{r^2}{n^2}\right)^{-n-(1/2)} \left(1 - \frac{r}{n}\right)^{-r} \left(1 + \frac{r}{n}\right)^{r}$$

$$= \left(1 - \frac{r^2}{n^2}\right)^{-(n^2/r^2)(r^2/n)} \left(1 - \frac{r^2}{n^2}\right)^{-1/2} \left(1 + \frac{r}{n}\right)^{n/r(-r^2/n)} \left(1 - \frac{r}{n}\right)^{-n/r(-r^2/n)}$$

Thus expression (A1–1) can now be written

$$\frac{1}{\sqrt{n\pi}} \left(1 - \frac{r^2}{n^2}\right)^{-(n^2/r^2)(r^2/n)} \left(1 - \frac{r^2}{n^2}\right)^{-1/2} \left(1 + \frac{r}{n}\right)^{n/r(-r^2/n)} \left(1 - \frac{r}{n}\right)^{-n/r(-r^2/n)}$$

Now

$$\left(1 - \frac{r^2}{n^2}\right)^{-n^2/r^2} \longrightarrow e \quad \text{as} \quad \frac{n}{r} \longrightarrow \infty$$

$$\left(1 - \frac{r^2}{n^2}\right)^{-1/2} \longrightarrow 1$$

$$\left(1 + \frac{r}{n}\right)^{n/r} \longrightarrow e$$

$$\left(1 - \frac{r}{n}\right)^{-n/r} \longrightarrow e$$

Thus finally, the probability of error R is

$$\frac{1}{\sqrt{\pi n}} e^{-r^2/n}$$

The significant feature of this result is the form e^{-r^2}. It specifies the probability of an error R and is thus equivalent to Eq. (3–3), in which the error is the difference between the true value X and the measured value x. The only problem that remains in putting the equation into standard form is to redefine the constants. Put

$$hx = \frac{r}{\sqrt{n}}$$

for the exponent and in the constant replace $1/\sqrt{n}$ by $h\, dx$. The equation then reads

$$P(x)\, dx = \frac{h}{\sqrt{\pi}} e^{-h^2 x^2}\, dx$$

where $P(x)\, dx$ is the probability of finding an error between x and $x + dx$.

A1–2 STANDARD DEVIATION OF THE GAUSSIAN DISTRIBUTION

We must calculate the sum of the squares of the errors divided by the total number of observations. Let there be N observations, where N can be assumed to be a very large number. The number of errors of magnitude between x and $x + dx$ equals $(Nh/\sqrt{\pi})e^{-h^2 x^2}\, dx$.

Therefore,

$$\sigma^2 = \frac{1}{N} \int_{-\infty}^{\infty} N \frac{h}{\sqrt{\pi}} e^{-h^2 x^2} \cdot x^2\, dx$$

$$= \frac{h}{\sqrt{\pi}} \int_{-\infty}^{\infty} x^2 e^{-h^2 x^2}\, dx$$

The integral is a standard one and has a value $\sqrt{\pi}/2h^3$. Therefore,

$$\sigma^2 = \frac{h}{\sqrt{\pi}} \frac{\sqrt{\pi}}{2h^3} = \frac{1}{2h^2}$$

This provides the justification for Eq. (3–4) and enables us to rewrite the probability function

$$P(x)\, dx = \frac{1}{\sqrt{2\pi}\, \sigma} e^{-x^2/2\sigma^2}\, dx$$

A1–3 AREAS UNDER THE GAUSSIAN DISTRIBUTION CURVE

The probability that an error falls between x and $x + dx$ is

$$\frac{1}{\sqrt{2\pi}\,\sigma}e^{-x^2/2\sigma}\,dx$$

Therefore, the probability that an error lies between 0 and x is

$$\int_0^x \frac{1}{\sqrt{2\pi}\,\sigma}e^{-x^2/2\sigma}\,dx$$

Although this integral can be easily evaluated for infinite limits, it is not so simple for fixed limits as we now require.

Numerical methods of integration are used with results given in Table A1–1 (see Fig. A1–2).

TABLE A1–1   Areas under the Gaussian Curve

x/σ	Probability that an error lies between 0 and x
0	0
0.1	0.04
0.2	0.08
0.3	0.12
0.4	0.16
0.5	0.19
0.6	0.23
0.7	0.26
0.8	0.29
0.9	0.32
1.0	0.34
1.1	0.36
1.2	0.38
1.3	0.40
1.4	0.42
1.5	0.43
1.6	0.45
1.7	0.46
1.8	0.46
1.9	0.47
2.0	0.48
2.5	0.49
3.0	0.499

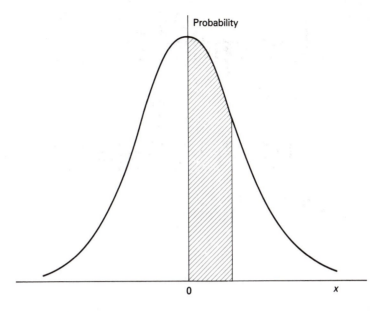

Figure A1–2 The area evaluated in calculating the probability of occurrence of an error up to x.

If we require the probability that an error lies between $\pm x/\sigma$, the value is, of course, doubled. For example, the entry at $x/\sigma = 1$ is 0.34, giving the 68% figure that we have been using for $\pm 1\sigma$ limits. The table is intended to give only an indication of the way the probabilities run, and for statistical work reference should be made to one of the many statistical tables available. (See the Bibliography under Lindley and Miller.)

Appendix 2

The Principle of Least Squares

A2–1 LEAST SQUARES AND SAMPLE MEANS

Consider that we make N measurements, x_i, of a quantity that contains random fluctuation. Let us calculate that value, X, whose deviations from the x_i are minimized in accordance with the principle of least squares. X will be obtained from the condition

$$\sum (x_i - X)^2 = \text{minimum}$$

Let $\bar{x}$ be the mean of the x_i. Then

$$\sum (x_i - X)^2 = \sum [(x_i - \bar{x}) + (\bar{x} - X)]^2$$

$$= \sum [(x_i - \bar{x})^2 + (\bar{x} - X)^2 + 2(x_i - \bar{x})(\bar{x} - X)]$$

or, since $\sum (x_i - \bar{x}) = 0$,

$$\sum (x_i - X)^2 = \sum [(x_i - \bar{x})^2 + (\bar{x} - X)^2]$$

This last expression clearly has a minimum value when $\bar{x} = X$, thus confirming that the use of the mean as the most probable value for a sample is consistent with the principle of least squares.

A2–2 LEAST-SQUARES FIT TO STRAIGHT LINES

Consider a set of observations (x_i, y_i) to which it is desired to fit a linear relation

$$y = mx + b$$

We assume that the x values are precise, that all the uncertainty is contained in the y values, and that the weights of the y values are equal. The differences whose sum of squares we desire to minimize are of the form

$$\delta y_i = y_i - (mx_i + b)$$

Therefore,

$$(\delta y_i)^2 = [y_i - (mx_i + b)]^2$$
$$= y_i^2 + m^2 x_i^2 + b^2 + 2mx_i b - 2mx_i y_i - 2y_i b$$

If there are n pairs of observations, the sum is

$$M = \sum (\delta y_i)^2 = \sum y_i^2 + m^2 \sum x_i^2 + nb^2 + 2mb \sum x_i$$

$$- 2m \sum x_i y_i - 2b \sum y_i$$

The condition for the best choice of m and b is that $\sum (\delta y_i)^2$ should be a minimum. We need, therefore,

$$\frac{\partial M}{\partial m} = 0 \quad \text{and} \quad \frac{\partial M}{\partial b} = 0$$

The first condition gives

$$2m \sum x_i^2 + 2b \sum x_i - 2 \sum (x_i y_i) = 0$$

the second

$$2nb + 2m \sum x_i - 2 \sum y_i = 0$$

Solution of the simultaneous equations for m and b gives

$$m = \frac{n \sum (x_i y_i) - \sum x_i \sum y_i}{n \sum x_i^2 - (\sum x_i)^2}$$

$$b = \frac{\sum x_i^2 \sum y_i - \sum x_i \sum (x_i y_i)}{n \sum x_i^2 - (\sum x_i)^2}$$

Standard deviations for m and b can be calculated as follows. The calculated values of m and b are functions of the quantities y_i. The standard deviations for m and b will, therefore, be calculated using Eq. (3–8) for the standard deviation of computed functions. They will be calculated in terms of the standard deviation of the y values. This was written as Eq. (6–5) using the quantities δy_i:

$$s_y = \sqrt{\frac{\Sigma\,(\delta y_i)^2}{n - 2}}$$

Justification of the value $n - 2$ will not be attempted; it is associated with the fact that the δy_i are not independent but are connected by the existence of the best line specified by the values of m and b. Equation (3–8) for the standard deviation of a computed value is

$$s^2 = \left(\frac{\partial f}{\partial x}\right)^2 s_x^2 + \left(\frac{\partial f}{\partial y}\right)^2 s_y^2 + \cdots$$

We apply this formula to our case by noting that the x and y of the formula are the y_1, y_2, etc., which form part of our set of observations. Thus the function for m is

$$m = \frac{1}{n\,\Sigma\,x_i^2 - (\Sigma\,x_i)^2}\left[nx_1 y_1 - y_1 \Sigma\,x_i + nx_2 y_2 - y_2 \Sigma\,x_i + \cdots\right]$$

Therefore,

$$\frac{\partial m}{\partial y_k} = \frac{1}{n\,\Sigma\,x_i^2 - (\Sigma\,x_i)^2}\left[nx_k - \Sigma\,x_i\right]$$

and

$$\left(\frac{\partial m}{\partial y_k}\right)^2 = \frac{1}{(n\,\Sigma\,x_i^2 - (\Sigma\,x_i)^2)^2}\left[n^2 x_k^2 + \left(\Sigma\,x_i\right)^2 - 2nx_k \Sigma\,x_i\right]$$

Since x_y is common to all the contributions, we can sum the $(\partial m/\partial y_k)^2$ directly to obtain

$$\sum_k \left(\frac{\partial m}{\partial y_k}\right)^2 = \frac{1}{(n\,\Sigma\,x_i^2 - (\Sigma\,x_i)^2)^2}\left[n^2 \Sigma\,x_i^2 + n\left(\Sigma\,x_i\right)^2 - 2n\left(\Sigma\,x_i\right)^2\right]$$

since $\Sigma\,x_k = \Sigma\,x_i$, etc.

Therefore,

$$\sum_k \left(\frac{\partial m}{\partial y_k}\right)^2 = \frac{1}{(n \sum x_i^2 - (\sum x_i)^2)^2}\left[n^2 \sum x_i^2 - n\left(\sum x_i\right)^2\right]$$

$$= \frac{n}{n \sum x_i^2 - (\sum x_i)^2}$$

or

$$s_m = s_y \sqrt{\frac{n}{n \sum x_i^2 - (\sum x_i)^2}}$$

The value for s_b can be found by the same procedure.

A2–3 WEIGHTING IN STATISTICAL CALCULATIONS

When we perform some statistical calculation, such as to obtain the mean of a set of observations or fit a function to observations using the least-squares principle, the equations in Secs. 3–3 and 6–7 are valid only when all the observations are equally precise. If the measurements are of unequal precision, it is obviously fallacious to allow them to make equal contributions toward the final answer. Clearly the more precise measurements should play a more important part in the calculation than the less precise values. In order to accomplish this we assign to the observations "weights" that are inversely proportional to the standard deviations of the observations. The derivations of the resulting equations will be found in the standard texts on statistics, and we shall simply quote the results here.

(a) Weighted Mean of a Set of Observations

Consider that we have a set of independently measured quantities, x_i, and that we know the standard deviations, S_i, for each of the x_i. The weighted mean of the set of values of x is given by

$$\bar{x} = \frac{\sum \dfrac{x_i}{S_i^2}}{\sum \dfrac{1}{S_i^2}}$$

and the standard deviation of a weighted mean by

$$S^2 = \frac{\sum \dfrac{(x_i - \bar{x})^2}{S_i^2}}{(N - 1) \sum \dfrac{1}{S_i^2}}$$

(b) Straight-Line Fitting by Weighted Least Squares

Consider that we have a set of values of a variable y measured as a function of x. As in Sec. 6–7 we shall assume that the x values are precise and that all the uncertainty is confined to the y values. The equations by which we can calculate the slope m and the intercept b of the best line can be written as follows:

$$m = \frac{\Sigma\, wy \, \Sigma\, wx^2 - \Sigma\, wx \, \Sigma\, wxy}{\Sigma\, w \, \Sigma\, wx^2 - (\Sigma\, wx)^2}$$

$$b = \frac{\Sigma\, w \, \Sigma\, wxy - \Sigma\, wx \, \Sigma\, wy}{\Sigma\, w \, \Sigma\, wx^2 - (\Sigma\, wx)^2}$$

Because of the cumbersome nature of these equations, we have used abbreviated notation in which we have dropped the suffix i that should be attached to each of the quantities. Also the term w_i is used for the weight of each pair of measured values (x_i, y_i). The weights will be calculated in terms of the standard deviations of the y values as

$$w_i = \frac{1}{(S_{y_i})^2}$$

The best estimates of the standard deviations for m and b can be written (as they were in Sec. 6–7) in terms of the deviations of the measured points from the best line. For a weighted least-squares fit these deviations must now be weighted, and the weighted value of S_y is given by

$$S_y = \sqrt{\frac{\Sigma\, w_i \delta_i^2}{n - 2}}$$

The best estimates of the standard deviation for the slope and intercept can now be written

$$S_m^2 = S_y^2 / W$$

$$S_b^2 = S_y^2 \left(\frac{1}{\Sigma\, w} + \frac{x^2}{W} \right)$$

where

$$W = \Sigma\, w(x - \bar{x})^2$$

and $\bar{x}$ is the weighted mean of the x values, given by

$$\bar{x} = \frac{\Sigma\, wx}{\Sigma\, w}$$

The suffix i has, as before, been omitted.

Appendix 3

Difference Tables and the Calculus of Finite Differences

A3–1 MATHEMATICAL FOUNDATIONS

The calculus of finite differences supplies a powerful tool for the treatment of measured variables. For the moment, however, let us consider the situation wholly from the mathematical point of view. After we have established mathematically the results we need, we can proceed to apply them to measurements.

Consider a known function $y = f(x)$ (see Fig. A3–1), which can be expressed in terms of a Taylor expansion about the value $x = a$:

$$f(x) = f(a) + (x - a)\left(\frac{df}{dx}\right)_a + \frac{(x - a)^2}{2!}\left(\frac{d^2f}{dx^2}\right)_a + \cdots$$

Such a function is said to be analytic at the point $x = a$, and any good book on calculus will provide more detail.

Such a function is defined along the continuous range of values on the scale of x, and, in order to make the theory applicable to measured variables, we must convert the formulation so that it refers to discrete values of x. Let these discrete values of x be spaced equidistantly upward from $x = a$ at intervals of h, so that the values of x in which we are interested are

$$x = a, \quad x = a + h, \quad x = a + 2h, \quad x = a + 3h, \ldots$$

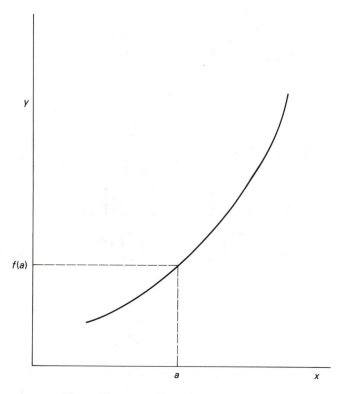

Figure A3–1 Graph of the function $y = f(x)$.

We can now calculate corresponding values of y for these discrete values of x. They will be

$$f(a), \quad f(a + h), \quad f(a + 2h), \quad f(a + 3h), \ldots$$

and we can illustrate these values on a graph as shown in Fig. A3–2.

If we concentrate our attention on these discrete values of x and y, and, if we wish to find a form of the Taylor expansion applicable to the discrete values, we can simulate the required derivatives using finite differences. We define the quantity $\Delta f(a)$ to be

$$\Delta f(a) = f(a + h) - f(a)$$

Correspondingly, we have

$$\Delta f(a + h) = f(a + 2h) - f(a + h)$$

etc., and these quantities will be related to the first derivatives of the function

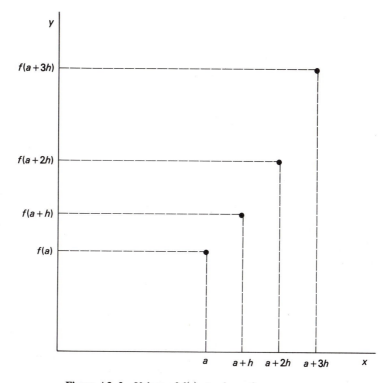

Figure A3–2   Values of $f(x)$ at values of x near $x = a$.

at the various values of x. In a similar way we define the second differences,

$$\Delta^2 f(a) = \Delta f(a + h) - \Delta f(a)$$

and so on for third and higher differences.

When we arrange all these differences beside the tabulated values of $f(x)$, we obtain a "difference table" for the values. A difference table for the function

$$y = 2x + x^3$$

is shown in Table A3–1. It illustrates several important features of difference tables, including, for this example, the constancy of the third differences and the consequently zero value of the fourth differences.

Now let us consider the problem of obtaining values of y at values of x intermediate between the discrete values of x. Furthermore let us find a way of

TABLE A3–1   Difference Table for the Function $y = 2x + x^3$

x	y	Δ	Δ^2	Δ^3	Δ^4
1	3				
		9			
2	12		12		
		21		6	
3	33		18		0
		39		6	
4	72		24		0
		63		6	
5	135		30		0
		93		6	
6	228		36		0
		129		6	
7	357		42		0
		171		6	
8	528		48		0
		219		6	
9	747		54		
		273			
10	1020				

calculating these intermediate values from the known values of y, instead of calculating them directly from the function itself. The advantage of such a procedure, of course, is that it will be available for later use on values for which we do not know the relevant function. In order to calculate these intermediate values we must rewrite the Taylor expansion in a form that is compatible with the quantities found in the difference table and also is suitable for the calculation of intermediate values. In Fig. A3–3 the gradient of the function f at $x = a$ can be approximated by the ratio Δ/h. Corresponding values for the second derivative can be calculated in terms of the second difference Δ^2, and so on for higher derivatives. Consider also a value of x intermediate between $x = a$ and $x = a + h$ and let it be specified by a new variable u that is defined by

$$x = a + uh$$

The Taylor expansion can now be rewritten in terms of the above mentioned quantities to yield the intermediate values of y as

$$y = f(a) + u\Delta + \frac{1}{2!}u(u-1)\Delta^2 + \frac{1}{3!}u(u-1)(u-2)\Delta^3 + \cdots$$

This form of the Taylor expansion is known as the Gregory-Newton formula for interpolation. It can be used to calculate intermediate values whenever we have tabulated values of two variables. For example, such methods used to be

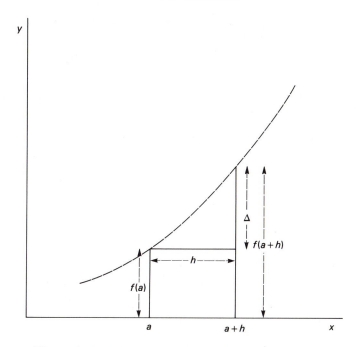

Figure A3–3   Approximation for the gradient of $f(x)$ at $x = a$.

commonly employed for interpolating in tabulated values of logarithms, trigonometric functions, etc.

To use the Gregory-Newton formula, we construct a difference table as far as those differences that become either zero or else small enough that the error involved in the interpolation calculation becomes acceptably small. If we construct the table so that the value we are seeking lies between $x = a$ and $x = a + h$, the various differences we need for insertion in the formula will be found along the upper edge of the table. If we are seeking a value that lies between $x = a + h$ and $x = a + 2h$, the differences we need will be found along the next lower row, and so on.

Extrapolation can be carried out by a similar process. Suppose we have a set of values of y for values of x ranging from $x = a$ to $x = a + (n - 1)h$. If we wish to calculate a value for y appropriate to $x = a + nh$, we must start with the supposition that the y values beyond $x = a + (n - 1)h$ are determined by the same function as for the lower values. On the basis of that assumption there is a simple method of finding y for $x = a + nh$; it is based on a process of extending the basic difference table. Starting with the column of differences that are constant, or sufficiently close to constancy for our purposes, we calculate successively the lower differences in terms of the higher until we reach the required value of y. (See Table A3–2.) In this way the table can be extended indefinitely to provide further values of the function as required.

TABLE A3–2   The Use of a Difference Table for Extrapolation

x	$y = x^3$	Δ	Δ^2	Δ^3
2	8			
		19		
3	27		18	
		37		6
4	64		24	
		61		6
5	125		30	
		91		6
6	216		30 + 6 = 36	
		91 + 36 = 127		
7	216 + 127 = 343			

We can also use a difference table to construct a polynomial that will either reproduce the actual functional relationship between y and x or else provide an adequate approximation thereto. To do this we must rewrite the Gregory-Newton formula in a form suitable for our purpose. We had written it in terms of the variable u; we now wish to write it in terms of the variable x while still incorporating the differences Δ rather than the derivatives df/dx. Remembering

$$x = a + uh$$

we have

$$u = \frac{x - a}{h}$$

and the original form of the Gregory-Newton equation becomes

$$y = f(x) = f(a) + \frac{1}{h}(x - a)\Delta + \frac{1}{2!}\frac{1}{h^2}(x - a)(x - a - 1)\Delta^2$$

$$+ \frac{1}{3!}\frac{1}{h^3}(x - a)(x - a - 1)(x - a - 2)\Delta^3 + \cdots$$

The equation is now in the form we desire. If we insert in it the appropriate values of Δ, Δ^2, Δ^3, etc., for a particular value such as $f(a)$, we shall generate the required polynomial in x.

As an example consider the difference table in Table A3–1 and choose the values in the top row. They are

$$f(a) = 3, \quad a = 1, \quad h = 1, \quad \Delta = 9, \quad \Delta^2 = 12, \quad \Delta^3 = 6, \quad \Delta^4 = 0$$

Inserting these values and performing some elementary algebra we obtain

$$y = 2x + x^3$$

Since this is the function we started with, we should not be surprised. However, we have confirmed the suitability of the Gregory-Newton formula for generating a polynomial that is consistent with a set of tabulated values. It is therefore of immense potential value for constructing a suitable polynomial when we are dealing with tabulated values alone and have no idea of a suitable function to act as a model.

A3–2 APPLICATION OF DIFFERENCE TABLES TO MEASURED VALUES

In the preceding section we have illustrated the calculus of finite differences and difference tables using mathematical functions. When we turn to measured variables and seek to apply these techniques, we encounter two differences: (a) we may not know a function that will provide an adequate fit to the observations, and (b) the variables will show scatter arising from uncertainty of measurement.

Consider first the case in which the measurements are very precise, so that the scatter is negligible in comparison with the measured values. In this case the difference table will contain values that behave relatively regularly, and we can use it to perform immediately such procedures as interpolation and extrapolation. Furthermore, if a polynomial of a certain order will serve as a good fit to the observations, the differences of the appropriate order will turn out to be nearly constant, and the next differences will scatter around zero. We can then use the procedures of the preceding section to construct the appropriate polynomial.

If, on the other hand, our observations show larger scatter, we are faced with a somewhat different problem of interpretation. It is in principle possible to fit a polynomial exactly to *any* set of values, no matter how much scatter they show. In fact to any set of values it is possible to fit an infinite number of polynomials, only two of which are illustrated in Figure A3–4. So which polynomial do we want? Is it going to be one like that represented by the solid line in Figure A3–4? Under some circumstances this may be appropriate. On many other occasions, however, we shall have good reason to believe that, measure-

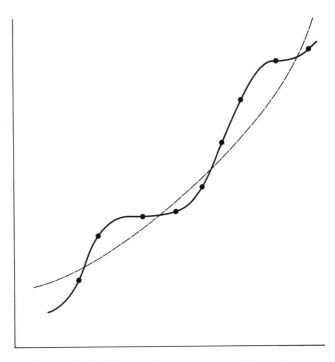

Figure A3-4 Fitting a polynomial to a set of observations.

ment uncertainty apart, the basic behavior of the system is regular, and we really want the function corresponding to the dotted line in Figure A3–4. We thus have to consider the topic of "smoothing" the observations, by which we mean choosing a function or curve that follows the observations in general terms but ignores deviations smaller than a selected limit. Many of the standard texts on measurement theory will supply detailed descriptions of smoothing procedures. See, for example, the text by Whittaker and Robinson listed in the Bibliography.

It is not always clear how far we have to go in smoothing observations. We trade off simplicity of representation against possible loss of genuine detail in the behavior of the system. This takes good judgment on the part of the experimenter, and our decisions are not always greeted with universal approval. In any case, if we do want to make a choice of a certain order of polynomial to represent the observations, we can choose the corresponding difference column in the table to be constant and, on the basis of some averaged value of these differences, construct the polynomial we want.

If such a procedure is not to our liking and we are restricted to unavoidably noisy observations, our only alternative may be to use a least-squares procedure and thereby minimize the discrepancy between the observations and a function of a chosen type. Note, however, the important distinction between the use of the difference table and the least-squares procedure. The difference table will tell us the coefficients of a polynomial that is implied by the observations; the least-squares procedure will simply give us the optimized parameters of a function whose general form we must choose for ourselves.

Appendix 4

Specimen Experiment

A4–1 EXPERIMENT DESIGN

System

We are given a spring suspended from a stand, a pan to hold weights attached to the lower end of the spring, a set of weights, and a stopwatch with a scale divided into $\frac{1}{5}$ second.

Model

We are told that, on the assumption that the extension of a spring is proportional to the load on it, the period of oscillation, T, of a suspended load, m, given by

$$T = 2\pi\sqrt{\frac{m}{k}}$$

where k is a constant for a particular spring.

Requirement

We are asked to measure k for our spring with an uncertainty not greater than 10%.

Experiment Design

Following the steps listed in Sec. 5–3, we have:

(a) Our operating system will consist of the spring alone. We have been given no information about the pan for the weights, or any way of weighing it, so we must proceed without that knowledge.

(b) Our model contains only two variables, load m and period of oscillation T, so it is simple to decide that our input variable, the one we can control, will be m and the output variable will be T.

(c) To put the equation into straight-line form, our first idea might be to remove the square root. Squaring both sides of the equation gives us

$$T^2 = 4\pi^2 \frac{m}{k}$$

Comparison with the straight-line equation

$$\text{vertical variable} = \text{slope} \times \text{horizontal variable}$$

suggests that we could choose

$$\text{vertical variable} = T^2$$

$$\text{horizontal variable} = m$$

$$\text{slope} = \frac{4\pi^2}{k}$$

This is an acceptable choice, but the unknown, k, appears in the denominator of the slope. To simplify later arithmetic it is equally valid, and more convenient, to write the equation

$$m = \frac{k}{4\pi^2} T^2$$

where

$$\text{vertical variable} = m$$

$$\text{horizontal variable} = T^2$$

$$\text{slope} = \frac{k}{4\pi^2}$$

(d) In our case the range of the input variable, m, may be governed by the weights we have been given. In addition, however, we should consider the possibility of overloading the spring. Has anyone suggested, or is it written anywhere, that loads should be restricted? We might try a few weights on the pan to see how the spring behaves. One way or another we can choose a range of m that we feel comfortable with. The range of T values presents no problem, because it is determined by the system.

(e) Let us suppose that our weights are sufficiently precise that their uncertainty need not be considered. They are not totally precise, of course, and if we want to know what uncertainty they do have, we should look in the supplier's catalogue.

The only uncertainty, therefore, will arise from the timing measurements, and that uncertainty will depend on the precision with which we are capable of timing the oscillations. The only way to find that out is to try it. We choose a typical load, start the oscillations, and measure the time interval for, say, 10 oscillations. We must now decide what determines the uncertainty of the measurement. Is it the accuracy of reading the stopwatch, or is it our ability to watch the oscillations and start or stop the watch appropriately? Obviously we must test this by trying the measurement again, and we must continue to probe the measuring system in this way until we are sure we know our capabilities. We may decide, as in the present case, that we are sure that we can measure time intervals with an uncertainty of ± 0.3 sec.

This, however, does not complete our consideration of precision: we must evaluate the effect on our k values of this uncertainty in T. It is difficult to plan ahead very exactly, because we shall obtain our final value of k from the graph, but it is only prudent to check that our individual measurements have adequate precision to contribute significantly to the final result.

For example, suppose we timed a certain number of oscillations that gave a time interval, t, of 2 seconds. What would be the contribution of our ± 0.3 sec to the uncertainty in k? k is a function of t^2. Therefore

$$RU(k) = 2 \times \frac{0.3}{2} = 30\%$$

Clearly such a measurement will make little significant contribution to the determination of k with 10% precision. If we want 10% precision in k, we need 5% precision in t, and we can determine the consequent limits on our measuring process for t by arguing as follows:

If

$$RU(t) \text{ must not be greater than } 5\%$$

i.e.,

$$\frac{0.3}{t} \ngtr .05$$

then

$$t \nless \frac{0.3}{0.05} = 6 \text{ sec}$$

Thus, whatever the actual value of T, if we time a number of oscillations that give a total time interval of 6 seconds or more, the uncertainty in our timing measurements is likely to be acceptable. For convenience we might choose a constant number of oscillations for the various loads, but, if we were short of time in a long experiment, we could choose for each load the number of oscillations that gives us a satisfactory value of t.

(f) Let us decide, as a first guess, that we shall measure the time for 10 oscillations, knowing that even for the lowest load this gives us a measured time interval in excess of 6 sec, and that we shall time oscillations for loads of 0.1, 0.15, 0.2, 0.25, 0.3, 0.35, 0.4, 0.45, and 0.5 kg. Since we shall want to plot m vs. T^2 and incorporate the value of the absolute uncertainty in T^2, i.e., $2T \times AU(T)$, we should lay out the table that expresses our measurement program to have the headings:

Load, m, kg	Number of Oscillations	Time, t, sec	Period, T, sec	Period², T^2, sec²	$AU(T^2)$ sec²

Experimental Results

The next step is to make the actual measurements and fill in the table with the measured values of t and the values calculated for period and periods² with its absolute uncertainty. The result of this process will then appear as in the accompanying table.

Load, M, kg	Number of Oscillations	Time, t, sec	Period, T, sec	Period², T^2, sec²	$AU(T^2)$, sec²
0.10	10	8.2 ± 0.3	0.82 ± 0.03	0.67	±0.05
0.15	10	9.8 ± 0.3	0.98 ± 0.03	0.96	0.06
0.20	10	10.7 ± 0.3	1.07 ± 0.03	1.14	0.06
0.25	10	11.5 ± 0.3	1.15 ± 0.03	1.32	0.07
0.30	10	12.5 ± 0.3	1.25 ± 0.03	1.56	0.08
0.35	10	13.0 ± 0.3	1.30 ± 0.03	1.69	0.08
0.40	10	13.8 ± 0.3	1.38 ± 0.03	1.90	0.08
0.45	10	14.5 ± 0.3	1.45 ± 0.03	2.10	0.09
0.50	10	15.2 ± 0.3	1.52 ± 0.03	2.31	0.09

These values of m and T^2 must now graphed. Each plotted value must contain its range of uncertainty. m has no uncertainty and T^2 is uncertain by the amount listed in the final column, so that each "point" on the graph will be a little horizontal line.

Once the values are plotted, the graph will look as shown in Fig. A4–1. The next step is to interpret what we see in terms of the categories described in Sec. 6–4. We first observe the scatter of the points and consider if it is consis-

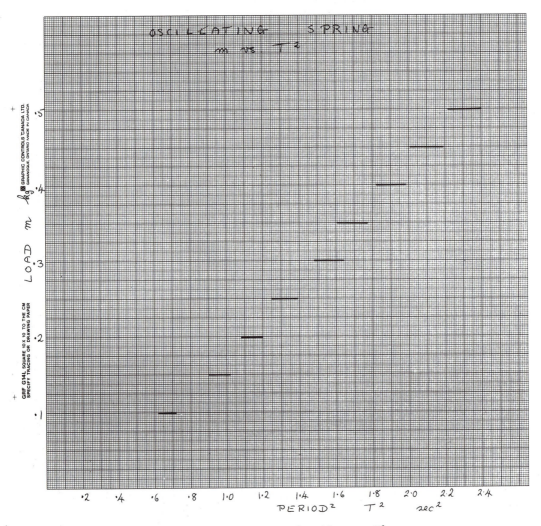

Figure A4–1   First stage of graphing m vs. T^2.

tent with our prior estimate of the uncertainty. In the present case there seems to be reasonable consistency between the scatter and the estimated uncertainty, and no further consideration of this point seems to be required at this stage. The next point to consider is the extent to which the behavior of the system is consistent with the model. In the present case the model predicted a straight line passing through the origin, and we must judge our graph against that. We can see immediately that in our case the correspondence with straight-line behavior seems quite adequate over the whole range. We shall be justified, therefore, in including all the points when we come to decide on our choice of lines.

With regard to intercept, however, the situation is different. A glance at the graph will make it clear that we are going to have an intercept on the T^2 axis that cannot be ascribed to measurement uncertainty. We shall have to return to consider this discrepancy later, but in the meantime we can note that the final value of k will be obtained from the slope alone, and that the slope can be calculated even in the presence of an unexpected intercept.

The next step is to draw lines so that we may obtain values for the slopes. One choice would be to draw our "best" line by eye and also lines that represent the maximum and minimum slopes permitted by the range of uncertainty in the scatter. At this stage the graph will look as shown in Fig. A4–2.

We now have to read values off the graph that will enable us to calculate these slopes. For each line we look for convenient intersections with the graph paper, illustrated in Fig. A4–3, that will give us the coordinates of points at the top and bottom of the line. On the present graph the chosen interactions are indicated by arrows, and the appropriate coordinates are marked. Given these coordinates, it is easy to calculate the slopes:

Steepest line:

$$\text{slope} = \frac{0.55 - 0.075}{2.40 - 0.73}$$

$$= 0.284 \text{ kg sec}^{-2}$$

Central line:

$$\text{slope} = \frac{0.55 - 0.05}{2.50 - 0.52}$$

$$= 0.253 \text{ kg sec}^{-2}$$

Least steep line:

$$\text{slope} = \frac{0.55 - 0.075}{2.64 - 0.51}$$

$$= 0.223 \text{ kg sec}^{-2}$$

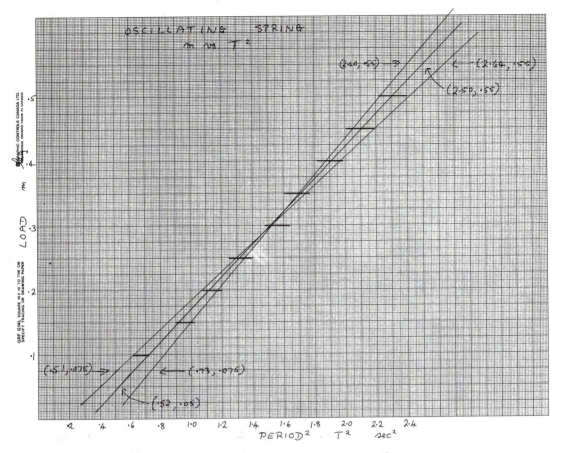

Figure A4–2 Final form of graph of m vs. T^2.

The corresponding values of k can now be calculated using

$$\text{slope} = \frac{k}{4\pi^2}$$

which gives

$$k = 4\pi^2 \times \text{slope}$$

Highest value:

$$k = 11.211 \text{ kg sec}^{-2}$$

Middle value:

$$k = 9.988 \text{ kg sec}^{-2}$$

Lowest value:

$$k = 8.804 \text{ kg sec}^{-2}$$

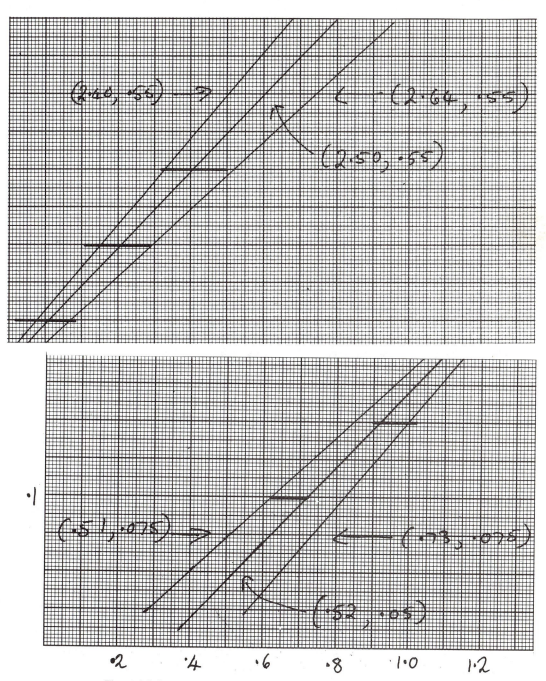

Figure A4–3 The identification of intersections at the top and at the bottom of the graph to permit calculation of the slopes.

Now that we have a measure of the overall uncertainty of the k values, we can round off the values to obtain our final statement about k and its uncertainty:

$$k = 10.0 \pm 1.2 \text{ kg sec}^{-2}$$

The final figure for the uncertainty at 12% of the k value is slightly greater than our target figure of 10%, but the difference is, perhaps, not too great for us to claim that we have come close enough to our objective. If any further reconsideration is forced on us, we could go back to the beginning and reassess our basic measurement uncertainty in timing. Certainly the low scatter of the points in comparison with the estimated uncertainty in the upper part of the graph suggests that we were slightly pessimistic about our estimate of ± 0.3 sec in timing, and reappraisal might allow us to refine that estimate.

Given the completion of our calculations for k, we now have to return to the question of the unexpected intercept. We satisfied ourselves that it was harmless because our k value was obtained from the slope, which could be determined accurately even in the presence of the intercept. Nevertheless we should not ignore it altogether, since it constitutes failure of correspondence between the model and the system, and it is not good experimental practice to leave things like that unconsidered. In guessing at possible sources of the discrepancy, we could note that it seems to be associated with some load not counted in our m values. For when our added load, m, is zero, the graph tells us that we would still observe a finite frequency of oscillation. What could give rise to such uncounted mass? One obvious guess would be the pan on which the weights were placed. Another obvious guess would be the mass of the spring itself. Without further investigation we cannot be certain that either of these guesses is good, but our explanation for the unexpected intercept seems reasonable enough that we are probably justified in terminating our present experiment at this point and leaving confirmation of our guesses to further experimenting.

A4–2 REPORT

In this section we shall give a version of a final report that could be written on the basis of the experiment whose conduct was described in the preceding section. The report will be written according to the suggestions offered in Chapter 7. Only the final version of the report is given; the details of its construction and their correspondence with the suggestions in Chapter 7 can be elucidated by comparing the report with the various sections of Chapter 7.

MEASUREMENT OF A SPRING CONSTANT BY AN OSCILLATION METHOD

Introduction

The stiffness of a spring can be measured by timing the oscillation of a suspended load. For an elastic spring (extension $\propto$ load) it can be proved that the period of oscillation, T, of a suspended mass, m, is given by

$$T = 2\pi \sqrt{\frac{m}{k}} \tag{1}$$

where k is a constant for a particular spring. The objective of the present experiment is to measure the value of k for a spring with an uncertainty not greater than 10%.

Equation (1) can be rewritten to read

$$m = \frac{k}{4\pi^2}T^2$$

which is linear in m and T^2 with slope $k/4\pi^2$. Thus by measuring the variation of oscillation period with load we shall be able to plot a graph of m vs. T^2 and obtain the value of k from the slope.

Procedure

Using the apparatus shown in Fig. 1, we measured the variation of oscillation period with load. The loads consisted of a Cenco weight set ranging from 0.1 kg to 0.5 kg with stated precision 0.04%.

The period of oscillation was measured by timing a number of oscillations using a stopwatch graduated in $\frac{1}{5}$ sec.

Results

The measurements of load and oscillation period are shown in Table 1.

The uncertainty shown for the measured times was estimated by visual judgment to be the outer limits for possible values of time, and so the calculated values for the uncertainty of T^2 also represent outer limits of possibility.

The graph of the m and T^2 values is shown in Fig. 2. The value of k and its uncertainty was calculated from the slopes of the three lines shown, giving

$$k = 10.0 \pm 1.2 \text{ kg sec}^{-2}$$

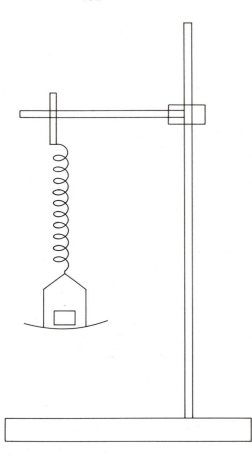

Figure 1   Diagram of spring, support, and load.

TABLE 1   Variation of Oscillation Period with Load

Load, M, kg	Number of Oscillations	Time, t, sec	Period, T, sec	Period^2, T^2, sec^2
0.1	10	8.2 ± 0.3	0.82 ± 0.03	0.67 ± 0.05
0.15	10	9.8 ± 0.3	0.98 ± 0.03	0.96 ± 0.06
0.20	10	10.7 ± 0.3	1.07 ± 0.03	1.14 ± 0.06
0.25	10	11.5 ± 0.3	1.15 ± 0.03	1.32 ± 0.07
0.30	10	12.5 ± 0.3	1.25 ± 0.03	1.56 ± 0.08
0.35	10	13.0 ± 0.3	1.30 ± 0.03	1.69 ± 0.08
0.40	10	13.8 ± 0.3	1.38 ± 0.03	1.90 ± 0.08
0.45	10	14.5 ± 0.3	1.45 ± 0.03	2.10 ± 0.09
0.50	10	15.2 ± 0.3	1.52 ± 0.03	2.31 ± 0.09

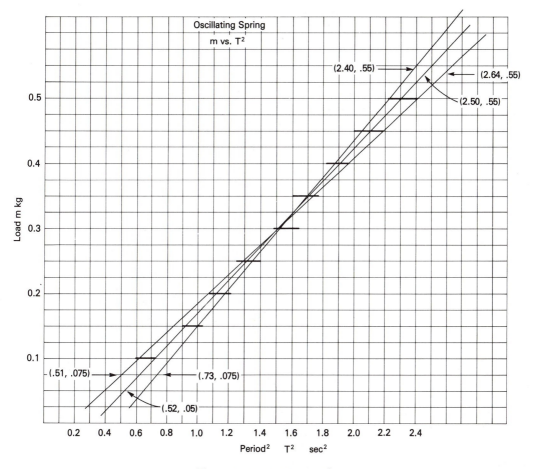

Figure 2 Graph of m vs. T^2.

Discussion

Using an oscillation method, we have measured the value of the stiffness constant for a spring to be $10.0 \pm 1.2 \text{ kg sec}^{-2}$. The model represented by equation (1) predicted for the m, T^2 graph a straight line passing through the origin. In our experiment the variation of m with T^2 proved to be consistent with a straight line within the present limits of uncertainty. Instead of passing through the origin, however, the lines shown in Fig. 2 can be seen to have a finite intercept on the T^2 axis that cannot be ascribed to measurement uncertainty. Our value of k, however, was obtained from the slope alone and should be free from error arising from factors that would give rise to an intercept.

Since the intercept gives a finite value for T at $m = 0$, it seems to be associated with the presence of some load not included in the measured values of m. Although we have not tested these possibilities, we can suggest that such extra load could arise from the pan supporting the weights and also the mass of the spring itself.

Bibliography

The subject of experimenting is vast, and the literature is correspondingly extensive and diverse. To meet all our varied needs there is no satisfactory substitute for going to the library and becoming familiar with its range of offerings. The following list contains some recent books and some classics, but it is very far from exhaustive. It is intended only to provide a few suggestions that may serve as a starting point for individual study.

BACON, R. H. "The 'Best' Straight Line among the Points," *American Journal of Physics,* **21,** 428 (1953).

BARFORD, N. C. *Experimental Measurements: Precision, Error and Truth*. Addison-Wesley, 1967.

BARRY, R. A. *Errors in Practical Measurement in Science, Engineering and Technology*. Wiley, 1978.

BEERS, Y. *Introduction to the Theory of Error*. Addison-Wesley, 1957.

BEVINGTON, P. R. *Data Reduction and Error Analysis for the Physical Sciences*. McGraw-Hill, 1969.

BRADDICK, H. J. J. *The Physics of the Experimental Method*. Chapman and Hall, 1956.

BRAGG, G. M. *Principles of Experimentation and Measurement*. Prentice-Hall, 1974.

BRINKWORTH, B. J. *An Introduction to Experimentation*. The English University Press, 1973.

COHEN, I. B. *Revolution in Science*. Harvard University Press, 1985.

COOK, N. H., RABINOWICZ, E. *Physical Measurement and Analysis*. Addison-Wesley, 1963.

COX, D. R. *Planning of Experiments*. Wiley, 1958.

DEMING, W. E. *Statistical Adjustment of Data*. Wiley, 1944.

DRAPER, N. R., SMITH, H. *Applied Regression Analysis*. Wiley, 1981.

FRETTER, W. B. *Introduction to Experimental Physics*. Prentice-Hall, 1954.

FREUND, J. E. *Modern Elementary Statistics*. Prentice-Hall, 1961.

HALL, C. W. *Errors in Experimentation*. Matrix Publishers, 1977.

HARRÉ, R. *Great Scientific Experiments*. Oxford University Press, 1983.

HORNBECK, R. W. *Numerical Methods*. Quantum Publishers, 1975.

JEFFREYS, H. *Scientific Inference*. Cambridge University Press, 1957.

KUHN, T. S. *The Structure of Scientific Revolutions*. University of Chicago Press, 1970.

LEAVER, R. H., THOMAS, T. R. *Analysis and Presentation of Experimental Results*. Macmillan, 1974.

LINDLEY, D. V., MILLER, J. C. P. *Cambridge Elementary Statistical Tables*. Cambridge University Press, 1958.

MARGENAU, H., MURPHY, G. M. *The Mathematics of Physics and Chemistry*. Van Nostrand, 1947.

MENZEL, D. H., JONES, H. M., BOYD, L. G. *Writing a Technical Paper*. McGraw-Hill, 1961.

PARRATT, L. G. *Probability and Experimental Errors in Science*. Wiley, 1961.

RABINOWICZ, E. *An Introduction to Experimentation*. Addison-Wesley, 1970.

ROSSINI, F. D. *Fundamental Measures and Constants for Science and Technology*. CRC Press, 1974.

SCHENCK, H. *Theories of Engineering Experimentation*. McGraw-Hill, 1961.

SHAMOS, M. H. (editor). *Great Experiments in Physics*. Holt-Dryden, 1960.

SHCHIGOLEV, B. M. *Mathematical Analysis of Observations*. American Elsevier, 1965.

SQUIRES, G. L. *Practical Physics*. McGraw-Hill, 1968.

STANTON, R. G. *Numerical Methods for Science and Engineering*. Prentice-Hall, 1961.

STRONG, J. *Procedures in Experimental Physics*. Prentice-Hall, 1938.

STRUNK, W., WHITE, E. B. *The Elements of Style*. Macmillan, 1979.

TAYLOR, J. R. *An Introduction to Error Analysis*. University Science Books, 1982.

TOPPING, J. *Errors of Observation and their Treatment*. The Institute of Physics, London, 1955.

TUTTLE, L., SATTERLEY, J. *The Theory of Measurements*. Longmans Green, 1925.

WHITTAKER, E. T., ROBINSON, G. *The Calculus of Observations*. Blackie, 1944.

WILSON, E. B. *An Introduction to Scientific Research*. McGraw-Hill, 1952.

WORTHING, A. G., GEFFNER, J. *Treatment of Experimental Data*. Wiley, 1946.

Answers to Problems

CHAPTER 2:

1. 142.45 ± 0.15 cm; 0.11%

2. 1.245 ± 0.005 A,
   3.3 ± 0.1V; 0.4%, 3.0%

3. 0.5 min

4. a) 10 cm,   b) 2 cm

5. 7.7%

6. 0.6%

7. 26 cm^2

8. 0.30 ohm   0.37   3.7%

9. 9.77 ± 0.04 m sec^{-2}, 0.4%

10. 1800 kg m^{-3}

11. 0.110 m, 0.0012 m, 1.08%

12. 0.8 nm, 0.24%

13. 14.3 ± 0.1; 14.25 ± 0.15

14. 6.75 ± 0.03

CHAPTER 3:

1.

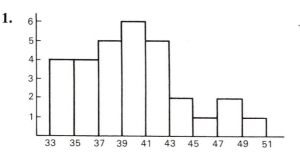

2. Between 39 and 41, 38

3. 38.3

4. 4.39

5. 0.80

6. 0.58

7. a) 33.9 to 42.7,
   b) 29.5 to 47.1

8. a) 37.5 to 39.1,
   b) 36.7 to 39.9

9. a) 3.8 to 5.0,   b) 3.2 to 5.5

10. 0.16

11. Rejection

12. Sample 34, 47, 43, 40, 32 has mean 39.2 and standard deviation 5.56

   Sample 36, 40, 38, 43, 34 has mean 38.2 and standard deviation 3.12

13. More than 130

14. More than 200

15. 0.047 mm^2

16. 0.21×10^{-9} m

17. 0.11 m sec^{-2}

CHAPTER 5:

1. No

2. range $\propto \left(\dfrac{\text{velocity}}{g}\right)^2$

3. pressure $\propto \dfrac{\text{surface tension}}{\text{radius}}$

4. period $\propto \sqrt{\dfrac{\text{moment of inertia}}{\text{rigidity constant}}}$

5. deflection $\propto \left(\dfrac{\text{load force}}{y \times \text{radius}^2}\right)^a$
   $\times \left(\dfrac{\text{length}}{\text{radius}}\right)^b$ radius where a and b are arbitrary constants

6. s vertically, t^2 horizontally, slope is $\frac{1}{2}a$

7. T vertically, $n^2 l^2$ horizontally, slope is 4 m

8. P vertically, v^2 horizontally, slope is $\rho/2$

9. T^2 vertically, $\cos \alpha$ horizontally, slope is $4\pi^2 l/g$

10. d vertically, Wl^3 horizontally, slope is $4/Yab^3$

11. h vertically, $1/R$ horizontally, slope is $2\sigma/\rho g$

12. p vertically, T horizontally, slope is R/v

13. fv_0 vertically, $f - f_0$ horizontally, slope is v

14. l vertically, Δt horizontally, slope is $l_0\alpha$ and intercept is l_0, whence α

15. $\sin \theta_1$ vertically, $\sin \theta_2$ horizontally, slope is μ_2/μ_1

16. $1/s$ vertically, $1/s'$ horizontally, each intercept is $1/f$; or ss' vertically, $s + s'$ horizontally, slope is f

17. $1/c$ vertically, ω^2 horizontally, slope is L.

18. F vertically, $1/r^2$ horizontally, check for linearity

19. F vertically, $i_1 i_2/r^2$ horizontally, check for linearity and check F vs. i_1, F vs. i_2 and F vs. $1/r$ separately, holding other variables constant.

20. log Q vertically, t horizontally, slope is $-1/RC$

21. Z^2 vertically, $1/\omega^2$ horizontally, slope is $1/c^2$ and intercept is R^2

22. m^2 vertically, m^2v^2 horizontally, slope is $1/c^2$ and intercept is m_0^2

23. $1/\lambda$ vertically, $1/n^2$ horizontally, slope is $-R$, intercept is $R/4$.

CHAPTER 6:

1. c) 0.00129 ohm^2 sec^2

d) 0.00572 henry

e) 0.00145 ohm^2 sec^2, 0.00117 ohm^2 sec^2

f) The measured value of L can range from 0.00544 henry to 0.00606 henry

g) 6.00 ohm

h) The measured value of R can range from 5.39 ohm to 6.43 ohm.

i) $L = 0.0057 \pm 0.0004$ henry
$R = 6.0 \pm 0.6$ ohm

2. Mean is 17.54 and the standard deviation of the mean is 0.26

3. a) Slope is 0.0499 ohm deg^{-1} and the intercept is 11.92 ohm

b) $\alpha = 0.00419$ deg^{-1}

c) 0.0024 ohm deg^{-1}, 0.16 ohm

d) 0.00021 deg^{-1}

e) $\alpha = 0.00419 \pm 0.00021$ deg^{-1}
$R_0 = 11.92 \pm 0.16$ ohm

4. $a = 2.12$, $b = 2.98$

5. a) $i = 0.5\, e^{2v}$

b) $y = 0.6\, x^{2.4}$

c) $f = 6.2\, e^{-365/T}$

6. b) 0.73

7. 0.99

Index

Absolute uncertainty, 10

Calibration errors, 11
Comparison, between models and systems, 64*ff*, 108*ff*
Compensating errors, 21
Compound variables, plotting, 80
Computed quantities
  standard deviation of, 41
  uncertainty in, 12
Control group, 96
Correlation, 128
Correlation coefficient, 132
Curve fitting, by least squares, 121

Diagrams, in reports, 143
Differences
  in experiment design, 94
  standard deviation of, 44
  uncertainty in, 17
Difference tables, 165
Digital displays, uncertainty in, 9

Dimensional analysis, 90
Discussion, in reports, 147
Distribution, concept of, 26

Experiment design, 76*ff*
Exponential functions
  plotting, 125
  standard deviation of, 45
  uncertainty in, 15
Extrapolation
  graphical, 59
  using difference table, 168

Finite differences, calculus of, 163
Format, in reports, 139
Function finding, 123

Gaussian distribution, 30*ff*
  areas under, 32, 156
  equation, 31, 151
  standard deviation, 32, 155

Graphs
  drawing, 106
  linear, 69*ff*
  logarithmic, 83, 124
  in reports, 146
Gregory-Newton formula, 166

Histogram, 26

Input, 2
Input variable, 76
Intercepts
  by least squares, 120, 159
  standard deviation of, 121, 161
Interpolation
  graphical, 58
  using Gregory-Newton formula, 167
Introduction, in reports, 140

Least squares
  intercept, 120, 159
  non-linear functions, 121
  precautions, 122
  principle, 118
  sample means, 158
  slope, 120, 159
  standard deviations, 121, 160
  weighted, 121, 162
Logarithmic functions
  standard deviation of, 45
  uncertainty in, 15
Logarithmic plotting, 83, 124

Mean
  definition, 27
  weighted, 161
Measurement
  nature of, 8
  standards, 7
Median, 27
Mode, 27

Models
  concept, 50
  empirical, 56
  testing, 64*ff*
  theoretical, 62

Normal distribution (*see* Gaussian distribution)

Output, 2
Output variable, 76

Planning, of experiments, 84*ff*
Poisson distribution, 31
Polynomial fitting, using difference
  table, 168
Polynomial representation, 125
Population (*see* Universe)
Powers
  standard deviation of, 45
  uncertainty in, 14
Precautions, in reports, 143
Precision
  definition, 11
  of experiment, 86, 126
Procedure, in reports, 142
Products
  standard deviation of, 44
  uncertainty in, 19
Purpose, in reports, 141

Quotients
  uncertainty in, 20
  standard deviation of, 45

Rectification of equations, 78
Rejection of readings, 46
Relative uncertainty, 11
Reports
  diagrams, 143

discussion, 147
format, 139
graphs, 146
introduction, 140
precautions, 143
procedure, 142
purpose, 141
results, 144
significance of, 137
title, 138
Results, in reports, 144
Rounding-off, 9

Sample means
  distribution of, 35
  and least squares, 158
Sample standard deviations, distribution
  of, 36
Sampling, 34
Significant figures, 21, 128
Significant variables, 49
Slopes
  determination of, 115
  by least squares, 120, 159
  standard deviation of, 160
  uncertainty in, 116
Smooth curve, uses, 57*ff*
Smoothing, of observations, 170
Standard deviation
  best estimate of universe, 37
  of computed values, 41
  definition, 29
  of Gaussian distribution, 32, 155
  of intercept, 121, 161
  of the mean, 36
  of slope, 121, 160
  of the standard deviation, 39

Statistical analysis, of experimental
  quantities, 105
Statistical fluctuation, 24
Stirling's theorem, 152
Straight-line form, 78
Sums
  standard deviation of, 43
  uncertainty in, 17
System, definition, 2
Systematic error, 11

Title, in reports, 138
Trigonometric functions
  standard deviation of, 45
  uncertainty in, 15

Uncertainty
  absolute, 10
  in calculated quantities, 12
  definition, 10
  in experimental quantities, 104
  on graphs, 107
  relative, 11
Uncertainty calculations, general
  method, 18
Universe
  definition, 33
  mean, 34
  standard deviation, 34, 37

Weighting
  in least squares, 121, 162
  in means, 161